3D Printing in Space
太空中的3D打印

李　韵
李小军
沈大海　编著

化学工业出版社

·北京·

Springer

内容简介

本书聚焦太空探索领域一项重要且极具趣味性的技术——太空3D打印，系统阐述了这项创新技术的核心原理与应用场景。在各国相继公布探月计划的时代背景下，人类探索浩瀚宇宙的梦想正逐步成为现实，而3D打印技术为太空探索事业提供了革命性解决方案。全书配有100余幅彩色示意图，不仅以通俗易懂的方式阐释了基础理论与研究进展，更深入探讨了实现航天器回收、在月球表面打印房屋等具有现实意义的主题，同时针对深空探测任务中的设备故障应对策略提供了专业见解。作者团队基于在卫星载荷技术领域丰富的研发经验，将扎实的理论知识与工程实践相结合。

本书既可为科研人员及工程技术人员提供技术指引，也可作为高等院校航天相关专业的教学参考书，更是一把开启星辰大海征程的智慧钥匙。

图书在版编目（CIP）数据

太空中的3D打印 = 3D Printing in Space : 英文 / 李韵，李小军，沈大海编著. -- 北京：化学工业出版社，2025. 4. -- ISBN 978-7-122-46661-7

Ⅰ. V52；TB4

中国国家版本馆CIP数据核字第2024XZ1608号

本书由化学工业出版社与Springer Nature出版公司合作出版。版权由化学工业出版社所有。本版本仅限在中国内地（大陆）销售，不得销往中国台湾地区和中国香港、澳门特别行政区。

责任编辑：张海丽　　　　　　　　　　文字编辑：王　硕
责任校对：宋　夏　　　　　　　　　　装帧设计：刘丽华

出版发行：化学工业出版社
　　　　　（北京市东城区青年湖南街13号　邮政编码100011）
印　　装：北京建宏印刷有限公司
710mm×1000mm　1/16　印张10½　字数219千字
2025年7月北京第1版第1次印刷

购书咨询：010-64518888　　　　　售后服务：010-64518899
网　　址：http://www.cip.com.cn
凡购买本书，如有缺损质量问题，本社销售中心负责调换。

定　　价：68.00元　　　　　　　　　　　　版权所有　违者必究

Preface

3D Printing in Space has a wide range of application potential in the construction and maintenance of satellite payload, rockets, spacecraft in orbit throughout the life cycle, deep space exploration missions, etc. It is one of the key space technologies that many countries are vigorously developing. To certain extent, it affects the research progress of major aerospace projects such as lunar base construction, and deep space navigation and exploration, and is also one of the areas of focus for researchers and enthusiasts of aerospace technology.

3D printing, also known as additive manufacturing, is a manufacturing process that creates three-dimensional physical objects from digital model files by depositing raw materials layer by layer. It is an intuitive expression method from the virtual conceptual space to the real physical world. 3D printing is not only a cutting-edge technology, but also a mature idea with a history. As early as the 1940s, from the inception of the concept of 3D printing, scientists have been trying to use this intuitive means to pile up conceptual structures. But not until the 1980s, the world's first commercial 3D printer that combined computer drawing, solid-state lasers, and resin curing technology came out, breaking the shackles of 3D printing. Since then, with the use of various materials for printing and the invention of various printing processes including fused deposition, laser sintering, and photopolymerization, 3D printing can now manufacture complex structures from polymers, ceramics, metals to amorphous alloys.

3D printing is a revolutionary manufacturing technology. Because of its disruptive process and the huge potential to change current manufacturing processes, it has attracted widespread attention in all walks of life. It is hailed as one of the key technologies in Industry 4.0 era. In the future, humans will explore even broader spaces, and 3D printing is aiding the path of human space exploration.

For aerospace activities such as on-orbit work, space travel, deep space exploration, and planetary base construction, a series of problems such as scarce raw materials, high transportation costs, and long cycles makes space 3D printing a key means to break through the limitations of aerospace mission material shortages. 3D printing can help astronauts create objects on demand in space, such as replacement parts for repairs, custom equipment for scientific experiments, and even necessities for human settlement in space such as food and buildings, revolutionizing the way space is explored. At present, space agencies around the world are considering building 3D-printed planetary bases on the moon or even Mars in the near future.

This book uses a graphic explanation method to elaborate on the application and prospects of 3D printing technology in space. The book includes four parts and eight chapters. The first part, including Chaps. 1 and 2, mainly introduces the concept of 3D printing and the current mainstream 3D printing technology, explaining why to develop 3D printing technology in space, the difference from the ground, and the challenges faced. The second part, including Chaps. 3, 4 and 5, mainly introduces the state-of-the-art of in-orbit 3D printing technology and some application examples. The third part, including Chap. 6, mainly introduces the application of new materials and new technologies used in 3D printing in space. The fourth part, including Chaps. 7 and 8, describes the future development prospects of 3D printing in space and the technological concepts of 3D printing for material self-sufficiency in astronaut survival, medical care, and flight missions.

This book has also received guidance from several experts and scholars. Here, we would like to thank Guobao Feng, Shiwei Dong, Qi Li, Shuo Liu, and Yafeng Li from the China Academy of Space Technology (Xi'an) for their valuable suggestions and support. Thanks to Bin Chen and Yimeng Xiong from Xi'an Platinum Special Additive Technology Co., Ltd. for their valuable suggestions and some illustrations. Part of the work in this book was funded by the National Youth Top-notch Talent Project, the Shaanxi Province Youth Top-notch Talent Project, and the Natural Science Foundation Project (Project No. 12175176, 61901360).

3D Printing in Space is an emerging, cross-disciplinary, sub-technical field. In recent years, with the rapid development of computer technology, robotics technology, and intelligent manufacturing technology, various space powers have gradually planned the development blueprint of 3D printing. It is timely to write an easy-to-understand book, summarizing the current technical status and future prospects. This book refers to and quotes some related literature from global aerospace institutions. The completion of this book benefits from the conception and realization of these pioneering works, and we would like to express our deep gratitude.

Owing to the limitation of our knowledge, there must be mistakes and errors in the book. Your suggestions would be appreciated.

Yun Li
Xiaojun Li
Dahai Shen

About This Book

Exploring the vast universe has been a dream of mankind since its inception. 3D printing technology is one of the most important technological developments in recent years, regarded as "one of the major technological breakthroughs that will change human life and production in the 21st century". This book comprehensively introduces the definition, technical means, application scenarios, and future prospects of 3D printing in space.

How to achieve "waste recycling" in space ships? How to print "houses" on the lunar surface? What to do if the parts of the spacecraft fail during deep space exploration missions? Readers can find answers to these interesting questions in this book.

This book combines the engineering experience and theoretical foundation accumulated by the author and her research team in the field of satellite payload technology. It is rich in illustrations, easy to understand, and blends popular science and scientific research contents. Every reader who opens this book can use it as a fulcrum to leverage their own journey to the stars and the sea.

About the Authors

 **Yun Li** is Professor and engaged in research on satellite communication and novel space microwave technologies.

 **Xiaojun Li** is Professor and engaged in satellite communication technologies.

 **Dahai Shen** is Professor and engaged in satellite communication technologies.

Contents

Chapter 1 What is 3D Printing in Space 1
  1.1 What is 3D Printing 2
  1.2 Why We Print Satellites and Other Spacecraft 4
  1.3 What is the Difference Between 3D Printing in Space
      and on the Ground 7
  1.4 Challenges Faced by 3D Printing in Space 8
  1.5 Prospects of 3D Printing in Space 9

Chapter 2 Existing 3D Printing Technologies 11
  2.1 Fused Deposition Modeling Technology 11
  2.2 Electron Beam Freeform Fabrication Technology 13
  2.3 Selective Laser Sintering Technology 15
  2.4 Stereolithography Technology 17
  2.5 Concentrated Solar Power Technology 19

Chapter 3 3D Printing of Artificial Satellites and Rockets 21
  3.1 Materials Suitable for Use in Space 21
      3.1.1 ABS Plastic 22
      3.1.2 New High Polymer Compound Materials 24
      3.1.3 Metal Materials 25
      3.1.4 Piezoelectric Materials 29
  3.2 3D Printed Satellite Components 30
      3.2.1 3D Printed Filters 32
      3.2.2 3D Printed Antennas 34
      3.2.3 3D Printed Loop Heat Pipe 36
      3.2.4 3D Printed Atmospheric Spectrometer Mirror 37
      3.2.5 New Welding Technology Based on 3D Printing 38
  3.3 3D Printing of Satellite Structure 39
  3.4 3D Printed Rockets 43
      3.4.1 3D Printing Rocket Engine Components 44
      3.4.2 Existing 3D Printed Rockets 49

		3.4.3 3D Printed Rocket Launch and Landing Devices	50
3.5		3D Printed Engine Components	51
3.6		3D Printed Small Satellites	52

Chapter 4 On-Orbit 3D Printing ... 57
- 4.1 Challenges Faced by Spacecraft and Others in On-Orbit 3D Printing ... 59
- 4.2 Microgravity 3D Printing Equipment ... 61
- 4.3 On-Orbit 3D Printing of Composite Materials ... 62
- 4.4 On-Orbit 3D Printing of Satellite Components ... 64
- 4.5 On-Orbit Recycling of Materials ... 65
- 4.6 On-Orbit Construction and Assembly ... 67
  - 4.6.1 On-Orbit Construction 3D Printer Arm ... 67
  - 4.6.2 On-Orbit Construction of Large Space Trusses ... 68
  - 4.6.3 On-Orbit Repair of Large Space Equipment ... 72
- 4.7 On-Orbit 3D Printed Satellites ... 72

Chapter 5 3D Printing in Deep Space Exploration ... 75
- 5.1 Challenges Faced by 3D Printing of Lunar and Other Planetary Bases ... 76
- 5.2 Where Does the Raw Material Needed for Printing Planetary Bases Come From ... 77
- 5.3 Latest Developments in 3D Printing Technology for Lunar Bases in Various Countries ... 80
  - 5.3.1 China's Lunar Base Development Concept and 3D Printing Technology ... 80
  - 5.3.2 NASA's Lunar Base 3D Printing Technology ... 82
  - 5.3.3 European Space Agency's Lunar Base 3D Printing Technology ... 88
- 5.4 3D Printing in Mars Base ... 91
  - 5.4.1 Mars Life Health and Exploration Mission ... 91
  - 5.4.2 Mars Dune Alpha ... 92
  - 5.4.3 In-Situ 3D Printing of Mars Base ... 93
- 5.5 3D Printing in Deep Space Exploration Instruments ... 94
- 5.6 New Materials Suitable for Deep Space Exploration Under Extreme Temperature Environments ... 97

Chapter 6 New Materials for Space 3D Printing ... 101
- 6.1 3D Printing of Metallic Materials ... 101
  - 6.1.1 Heterogeneous Dual Metal 3D Printing ... 101
  - 6.1.2 3D Printing of Large Copper Parts ... 103
  - 6.1.3 3D Printing of Iron Nickel Metal ... 104
  - 6.1.4 3D Printing of Titanium Metal ... 106
  - 6.1.5 3D Printing of Metal Ceramic Parts ... 107
  - 6.1.6 Stainless Steel 3D Printing ... 107

6.2	3D Printing of Composite Materials		108
	6.2.1	On-Orbit Recycling and Reuse of 3D Printed Plastics	108
	6.2.2	Printing Bricks with Artificial Lunar Dust	109
	6.2.3	Printing Concrete with Urea and Artificial Lunar Dust	110
	6.2.4	Printing Filters with Silicon Carbide	112
6.3	3D Printing of Biomimetic Materials		112
	6.3.1	3D Printing of Bones and Skin	112
	6.3.2	3D Printing of Blood Vessels	113
	6.3.3	Emergency Bioprinter	113
	6.3.4	3D Printing Ensures Astronaut Life Systems	114
6.4	Status of Printing Materials		117
	6.4.1	Heat Resistance of 3D Printed Metal	117
	6.4.2	Laser 3D Printing X-ray Imaging	118
	6.4.3	Surface Condition of 3D Printing	118

Chapter 7 Technical Applications of Space 3D Printing 121

7.1	High-Temperature Resistant 3D Printing Technology		121
	7.1.1	3D Printing of High-Temperature Resistant Nozzles	121
	7.1.2	3D Printed Platinum Alloy Thruster Chamber	123
	7.1.3	3D Printed Insulation Technology	123
7.2	Complex Structure 3D Printing Technology		124
	7.2.1	3D Printed High Complexity Rocket Turbo-Pump	124
	7.2.2	3D Printing of Complex Aircraft Icing Shapes	125
	7.2.3	3D Printing of High-Precision Gas Identification Telescopes	127
	7.2.4	3D Printing of a Complete Storable Thrust Chamber	127
7.3	High Utilization of 3D Printing Technology		129
	7.3.1	3D Printing Aids in the Search for Black Holes	129
	7.3.2	Building a Lunar Base with 3D Printing	130
	7.3.3	3D Printing Helps Fight "COVID-19"	132
	7.3.4	3D Printing of Astronaut Medical Tools	134
7.4	Integrated 3D Printing Technology		135
	7.4.1	3D Printed Miniature Sensors	135
	7.4.2	3D Printed Miniature Satellite Body	136
7.5	Lightweight 3D Printing Technology		136
	7.5.1	Lightweight 3D Printed Components	137
	7.5.2	3D Printed Metal Microwave Devices	138
	7.5.3	Integrated 3D Printed Satellite Bracket	140

Chapter 8 Future Development of 3D Printing in Space 143

8.1	3D Printing of Food in Space	143
8.2	Resupply of Spaceship Parts During Interstellar Travel	144
8.3	What to Do if an Astronaut Gets Seriously Ill During Interstellar Travel	146
8.4	3D Printing of Space Landers	150

Main Sources of Information 153
Bibliography 155

Chapter 1
What is 3D Printing in Space

3D printing technology has made significant progress over the decades since its inception, and is even hailed as one of the new technologies that change traditional mechanical processing methods in the era of Industry 4.0, attracting widespread attention from various countries. For the aerospace industry, 3D printing technology has a positive role in promoting new manufacturing methods for satellites and rockets, intelligent manufacturing, on-orbit resupply, deep space exploration, etc., which will greatly reduce the development cost of spacecraft, shorten the development process, and even potentially solve problems that existing technologies cannot solve, producing disruptive innovative results.

Due to the outstanding advantages of 3D printing technology, such as fast forming speed, high material utilization rate, short production cycle, low cost of customized production, high degree of digitization, and conducive to intelligent manufacturing, China has launched the "National Additive Manufacturing Industry Development Promotion Plan" and formulated a scientific and technological development plan for 3D printing, leading in the manufacture of high-performance large parts through 3D printing technology; the United States has established the National 3D Printing Technology Innovation Center and released the "Space 3D Printing" report in the implementation of the advanced manufacturing powerhouse plan; the European Space Agency is also actively seeking breakthroughs in new technologies such as using 3D printing technology to build lunar bases and realize small satellite metal 3D printing.

3D printing technology in space shows vigorous vitality. With the participation of space agencies from various countries, preliminary development has been achieved in materials, equipment, printing technology, etc., laying the foundation for more extensive applications in the future.

The definition of 3D printing in space is quite broad. Any spacecraft operating in space, including satellites, spaceships, space stations, etc., as well as planetary bases built during the exploration of outer space, and even rockets that carry spacecraft, when they use 3D printing technology, we define it as 3D printing in space. This

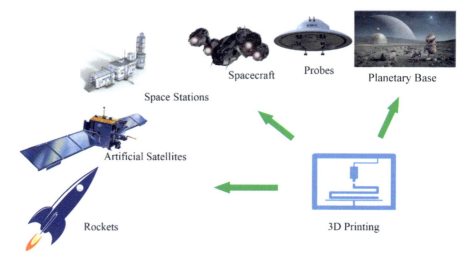

Fig. 1.1 Concept of 3D printing in space

includes both the technology used to produce and manufacture spacecraft on Earth and the technology that may be used for in-situ 3D printing in space (Fig. 1.1).

1.1 What is 3D Printing

3D printing, also known as additive manufacturing, is different from the traditional mechanical processing's subtractive manufacturing. Computer Numerical Control (CNC) is currently the mainstream production method in the manufacturing industry when high processing precision is required. On a given metal block, CNC mainly gives the processing route through a CNC machine tool, and then the cutter moves back and forth along the given route, completing turning, milling, etc., on the metal block until the processing is completed. The CNC machining quality is stable and the surface roughness is good, but to a certain extent, there is a large waste of raw materials, inability to process complex internal structures, and high cost of custom production. Traditional machining methods often obtain the designed product by gradually reducing a large piece of raw material, so it is aptly defined as "subtractive manufacturing".

3D printing is different, it can "use as needed" for raw materials, avoiding waste of raw materials. At the same time, it can complete complex external structures, hollow structures, especially the production with complex 3D internal configurations. Additive manufacturing, as the name suggests, is to obtain the designed product by accumulating raw materials bit by bit, which is different from subtractive manufacturing technology. Compared with other processing technologies, 3D printing technology

1.1 What is 3D Printing

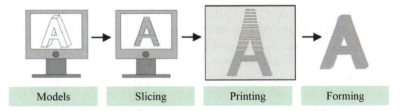

Fig. 1.2 3D printing process

has advantages such as flexible design, short processing cycle, low cost of customized design and implementation, and no need for additional splicing and assembly.

In our daily life, a printer can print two-dimensional data into flat graphics, obtaining photos that are colorful and almost identical to the data model. We can imagine 3D printing as "photo" printing in three-dimensional space. The difference is that we first need to establish a three-dimensional data model, and then use a 3D printer to "print" the three-dimensional data model layer by layer in real space to obtain the final desired three-dimensional object (Fig. 1.2). Just like the stacking process of Lego blocks, the "ink" sprayed by the 3D printer is equivalent to bricks, and finally stacks the shape we want. We often say "everything can be 3D printed", which means that the range of 3D printing is wide, including various daily necessities, industrial equipment, and even food and human organs.

The history of 3D printing technology can be tracked back to the photographic sculpture molding and terrain molding technology at the end of the nineteenth century. In many science fiction films, we can also see the prototype and shadow of 3D printers. But the birth of 3D printing technology in the true sense began with the development of the world's first commercial 3D printer—In 1986, Charles Hull in the United States obtained the first invention patent for a 3D printer that combined computer graphics, solid-state lasers, and resin curing technology. In the following ten years, people gradually developed various 3D printing technologies and corresponding 3D printers, and the materials available for 3D printing gradually enriched.

The most important innovation of 3D printing technology is to change the "design for manufacturing" in traditional manufacturing to "design for needs". In traditional processing and preparation, limited by processing technology, the designer needs to consider the feasibility of the design first in the design stage. For example, for microwave devices used in space, designers rarely consider designing them into a spherical shape. Because compared to the cylindrical shape, the spherical structure is less stable, and it cannot be formed in one go with traditional methods. Another example is the fuel tank of a spacecraft, which needs to use as little raw material as possible to accommodate as much raw material as possible under the greatest pressure. The best choice is obviously a sphere, but in actual use, a cubic container is often used, not only there are structural weaknesses at the corners and edges, but also the space cannot be maximized, which is also limited by existing processing technology. For some electronic devices, the connection point is often the most affected and the most vulnerable link. If engineers can integrate multiple components into a single manufacturing process, minimizing connections as much as possible, it will greatly benefit the stability and performance of the system. These are the advantages that 3D printing technology can bring to existing technology. Ideally, it allows designers to truly start from actual needs when designing, create products with better performance, and realize them through 3D printing.

The three elements of 3D printing are the 3D model, the printer, and the raw materials. The first step in achieving 3D printing is to have a 3D printer that can convert three-dimensional data into physical objects. Choosing the appropriate raw materials according to the design needs and the type of printer is the second step. The most crucial step is the creation of the 3D model. "There are a thousand Hamlets in a thousand people's eyes", similarly, when the preparation is no longer limited by the manufacturing process, the same preparation may give birth to countless possible designs. How to obtain the design scheme with the least raw materials, the most optimized structure, and the best performance is something that engineers need to pay attention to and consider when using 3D printing for design and creation.

1.2 Why We Print Satellites and Other Spacecraft

For the aerospace industry, 3D printing also opens a door of imagination for engineers. When product realization is no longer limited by the manufacturing process, each new solution corresponding to actual needs can be called a work of art, possibly combining innovation in function, material, and design.

In addition to creating special shapes with unique superior performance that traditional manufacturing processes cannot achieve, the application of 3D printing in space is far broader than we imagine.

For spacecraft manufactured and prepared on the ground, they are composed of various components and mechanical parts, and their structure and functions are very complex. For example, a typical communication satellite consists of antennas, receiver systems, digital signal processing systems, etc. For space stations, because they need to ensure astronauts' long-term living and working, they need to be equipped with more diverse and comprehensive devices. However, each device often requires a small quantity, and due to the harsh volume and weight restrictions in space, these devices often have special requirements different from ground applications. The aerospace industry is an industry of small batch production. For some components and materials, unlike the mass production of commercial products on the ground, only a few samples are often needed, but they are required to be well made. Traditional manufacturing processes reduce costs through mass production, while special customization of individual components is expensive, or for some manufacturing processes, such as metal injection molding technology, can produce high-precision, high-quality products, but often need to manufacture tens of thousands of products to reduce the cost of molds. Using 3D printing technology can overcome many problems faced by traditional manufacturing, and has advantages

1.2 Why We Print Satellites and Other Spacecraft

such as shorter processing procedures and lower costs for the customized design and implementation of individual components under specific application requirements and scenarios.

3D printing brings new technical support for the production of space application scenario models. For rocket development and simulation of space application scenarios, if a scaled-down model can be established first, it can give aerospace engineers a more intuitive understanding of the product and form a more accurate prediction of the design effect and performance. Such model production often has characteristics such as complex structure, high degree of refinement, and few production samples, and traditional manufacturing techniques are often limited by high costs and difficulty in processing. The process characteristics of 3D printing make it particularly suitable for the production of space models. First, 3D printing technology is easy to make samples of various complex structures and special structures (Fig. 1.3). Secondly, 3D printing technology is not limited by mass production molds, bringing new technical dawn to the production of models in space application scenarios (Fig. 1.4).

Fig. 1.3 3D printed astronaut suit parts. *Source* European Space Agency

Fig. 1.4 3D printed rocket model

Fig. 1.5 3D printed hollow dot matrix structure for weight reduction. *Source* European Space Agency

The aerospace industry's pursuit of reducing volume and weight is eternal. A decrease in volume and weight means a significant reduction in various costs, especially launch costs, and can even directly affect the success or failure of a spacecraft's launch mission. How to complete a design that meets performance requirements under stringent volume and weight requirements is one of the key issues troubling satellite system designers and technical designers. 3D printing technology brings new possibilities for product weight reduction design, mainly reflected in new structural design and implementation, weight reduction treatment of existing structures, and reduction of connection assembly points. In traditional manufacturing processes, components or devices often require additional support structures and connection structures, such as high-pressure connection devices in rocket engines, flanges in waveguide connections, and partitioning and connections in cavity structures, resulting in significant weight redundancy. Figure 1.5 shows a typical weight reduction design, which not only reduces the redundant structure at the connection point through integrated processing, but also can thin the cavity wall, and further use a hollow dot matrix structure on the existing structural parts to reduce the weight of the structural parts while achieving the required functions.

For spacecraft operating in space, 3D printing technology is key to the realization of many disruptive technologies. Existing spacecraft systems need to first consider factors that can withstand ground manufacturing, assembly, transportation, especially the load during the launch process, and the launch volume and weight are strictly limited by the launch vehicle. Carrying out 3D printing in space can eliminate these limiting factors, launch materials into space and then carry out microgravity 3D printing, realizing the innovative revolution of on-orbit 3D printing.

By deploying ultra-large space antennas on space operating orbits, it is possible to receive and capture weak cosmic signals. However, transport ultra-large space antennas to space and deploy them in orbit is very difficult, often meaning high launch costs, or even impossibility. In 2020, China listed "on-orbit 3D printing technology for ultra-large space antenna structures" as one of the top ten scientific and technological problems in the aerospace field of the year, and aerospace agencies in various countries are also actively exploring the implementation of large-scale in-orbit 3D printing technology. Achieving on-orbit 3D printing means that it is only necessary to complete the preparation of ultra-large space antennas in the space station, push them out of the cabin, and complete the deployment, or directly use the robotic arm to complete the printing outside the cabin, saving a lot of launch costs, turning "impossible" into "possible".

3D printing technology is also expected to provide solutions for the construction of lunar bases, Mars bases, and even bases on further planets. Once we can master the technical ability to continuously build bases on planets, we can provide habitats and information transfer support for astronauts and various equipment on the planet, making it possible for further resource exploration and scientific expeditions on the planet. 3D printing technology is not limited by the shape of the building, engineers can design base solutions that meet the specific environment of the planet. At the same time, combined with the technology of in-situ resource development and utilization for construction, it is possible to continuously build "bricks" until the construction of the planetary base is realized. Aerospace agencies and research institutions in various countries have launched scientific competitions and exploration activities for the construction of lunar bases, attracting a large number of researchers and interested students to participate.

In addition to this, 3D printing technology is indispensable for spacecraft in-orbit construction, deep space exploration component supply, and more. When it is impossible to obtain continuous and stable resource supplies from Earth, 3D printing technology will become a possible solution for astronauts encountering equipment and component failures, or for creating new tools or devices, making the future of interstellar voyages in people's imaginations possible.

1.3 What is the Difference Between 3D Printing in Space and on the Ground

3D printing in space differs significantly from 3D printing on the ground, from equipment and materials to energy supply.

The space environment is different from the ground on Earth, with complex and harsh environmental factors such as microgravity, high vacuum, strong radiation, space plasma, and extreme temperature changes. Materials and even 3D printers will face various challenges when operating in this environment for a long time, leading to reduced performance, and even possible loss of function.

3D printers in space not only need to adapt to the space environment such as microgravity, but also need to be as light and small as possible. This is not only to reduce launch costs, but also to save energy for long-term operation of space stations or spacecraft (which is often one of the decisive factors for the success or failure of long-term space flight). For ground-based 3D printers, especially industrial-grade 3D printers, they are often bulky, several meters long, and weigh hundreds of kilograms.

To achieve 3D printing in space, we first need to realize a 3D printer that can operate stably and for a long time under microgravity conditions in space missions, which is very necessary for stable material and equipment supply in orbit and deep space exploration. This is especially true for deep space exploration and the operation of bases on other planets, where the cost of obtaining round-trip cargo supplies from Earth is high and almost impossible to complete.

The materials used for 3D printing in space also differ from those on Earth. For example, for the construction of planetary bases, it is usually considered to use existing resources on the planet for in-situ construction. In spacecraft and on-orbit 3D printing, due to the limitations of material supply, existing materials and existing components are often recycled.

Corresponding to the model design of 3D printing in space, it also requires design under the premise of light weight and small volume, and additional consideration of specific application scenario requirements.

1.4 Challenges Faced by 3D Printing in Space

3D printing in space differs greatly from that on Earth and faces many challenges.

The most pressing challenge is the harsh environmental factors in space (Fig. 1.6). For on-orbit 3D printing, general 3D printers are placed inside space stations or spacecraft, shielding the extreme temperature environment in space, high-energy protons, space plasma, and cosmic rays, and the impact on the machine can be ignored. However, for products printed by 3D printing, such as space antennas and small satellites working outside the cabin, achieving long-term and stable on-orbit operation is a major challenge due to reliability issues caused by the space environment.

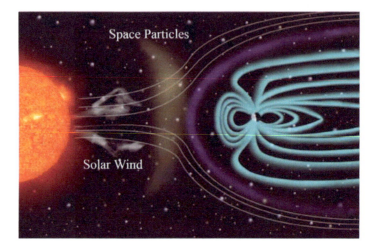

Fig. 1.6 Space environment

The microgravity or even zero-gravity environment in space will significantly affect the processing accuracy of 3D printers, the solidification process of materials, the fixation method of equipment, and the stepping method of position, which will affect the mechanical properties and functionality of the printed products. In a microgravity environment, gravity is only one ten-thousandth to one millionth of that on Earth, posing a huge challenge to the development of new printing processes and printers, requiring further on-orbit verification.

3D printing in space often takes place with limited human involvement or under unmanned operation/unattended conditions. How to transmit data such as models, remote control commands, and status monitoring, is a problem. And establish a stable equipment-ground communication link, is a prerequisite for stable 3D printing in space, especially for intelligent manufacturing in space.

For the 3D printing construction of space bases, the difficulties are multifaceted. On the one hand, there is a lack of sufficient raw materials. It is expensive and impractical to transport raw materials from Earth. Although many planets have construction raw materials such as dust and rocks, the composition of these raw materials is obviously different from that on Earth. It is necessary to adapt to local conditions and develop methods for preparing construction materials suitable for that planet. On the other hand, the printer also needs to be able to operate normally under extreme operation conditions on the planet. Factors including microgravity, dust, radiation, extreme temperature changes, and etc., may make machines that operate normally on Earth to be "unacclimated" on the planet and even destroy. In addition, the autonomous operation of the printer, remote control, and sustainable energy supply are also problems to be solved.

1.5 Prospects of 3D Printing in Space

3D printing in space is gradually developing from ground construction and implementation verification in space stations to lunar construction and Mars construction, and will show a variety of application prospects in future human space exploration activities (Fig. 1.7).

The more comprehensive the research on Earth, the more conducive it is to carry out preliminary technical research and verification on parts printed in space, manufactured equipment, constructed platforms, etc., so as to fully assess their feasibility and risk.

At present, aerospace engineers have developed some printable structures, electronic devices, and spacecraft parts (Fig. 1.8).

In space stations, the effects of microgravity and high vacuum on the construction process and the performance of the products formed can be fully evaluated. At present, countries such as China and the United States have carried out certain on-orbit 3D printing work in space stations, providing experimental data and a basis for subsequent on-orbit construction of microsatellites, on-orbit construction of large space devices, and recycling of materials in space.

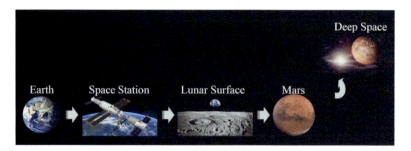

Fig. 1.7 Development history of 3D printing in space

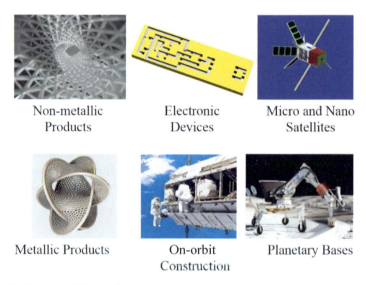

Fig. 1.8 Applications of 3D printing in space

Building on the moon and Mars using 3D printing has also received widespread attention from major aerospace institutions and scientists around the world. The conduct of these scientific research activities will draw a beautiful blueprint for human deep space exploration and long-distance deep space navigation.

3D printing in space is still in the early exploration stage, and there are still many technical difficulties in the conduct of space activities such as spacecraft preparation and large-scale space construction. Some fields have technical gaps and need further exploration to ensure that humans could subsequently carry out high-quality space operations, high-end aerospace missions, and deep space exploration.

Chapter 2
Existing 3D Printing Technologies

3D printers are the most important carriers for realizing 3D printing.

Correspondingly, a variety of 3D printing technologies have been developed in recent years, capable of printing different materials, models, and structures. According to different material properties, 3D printers can be divided into metal 3D printers, non-metal 3D printers, special material 3D printers, etc. And based on different printing molding technologies, 3D printing technologies can be simply divided into fused deposition modeling technology, laser sintering technology, photopolymerization molding technology, etc.

2.1 Fused Deposition Modeling Technology

The working principle of Fused Deposition Modeling (FDM) technology can be summarized as follows.

The material is heated and melted, extruded from the nozzle in a certain size. Then the nozzle moves on the workbench plane according to the path preset by the software, forming a layer of graphics. After that, the workbench descends a layer thickness in the vertical direction, the nozzle continues to print, finally forming the designed three-dimensional model (Fig. 2.1).

The advantage of fused deposition modeling technology is that its working principle is relatively simple, it does not require lasers and other more expensive components, and it is easier to maintain and operate. It has lower requirements for the working environment, making it an ideal candidate for desktop 3D printers. The model structure formed by printing has higher strength and stability, which can meet the mechanical performance requirements of most applications. Its main disadvantage is that the printing speed is relatively slow, which is not conducive to printing large models. The printing accuracy is relatively low compared to other 3D printing technologies (the printing accuracy of existing industrial-grade FDM printers can

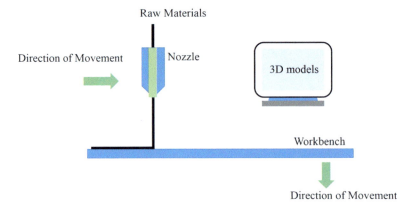

Fig. 2.1 Working principle of fused deposition modeling (FDM) technology

only reach the level of hundreds of micrometers, which is 0.1 mm in precision). At the same time, the surface of the printed model is relatively rough, and there is a step effect when observed under magnification, which is not suitable for applications that require high surface roughness. Fused deposition modeling technology is currently widely used in communications, electronics, automobiles, medicine, construction, and even toys, home appliances industries (Fig. 2.2).

In 2014, the world's first space 3D printer sent to the International Space Station by NASA used fused deposition modeling technology. It was a desktop FDM printer (Fig. 2.3). This 3D printer was developed by the American Space Manufacturing Company and can prepare some relatively simple common plastic tools and devices in space. The printing raw material is thermoplastic plastic—acrylonitrile–butadiene–styrene (ABS plastic). The ABS plastic is first heated to the transition temperature, then sprayed onto the workbench through the nozzle, and the solidification time of the plastic is almost instantaneous. The external dimensions of the printer are approximately 33 cm × 30 cm × 36 cm.

Fig. 2.2 FDM printer and printed model

2.2 Electron Beam Freeform Fabrication Technology

Fig. 2.3 The first space 3D printer sent to the International Space Station by NASA

Fig. 2.4 The space 3D printer sent to the space station by the European Space Agency and its printed antenna support structure. *Source* European Space Agency

In 2015, the European Space Agency sent a FDM printer developed by the Italian company Altran to the space station to conduct a verification test of 3D printing in space. The printing material is biodegradable material or non-toxic plastic (Fig. 2.4). Altran named it the "POP3D Portable On-board Printer", this FDM printer also has a desktop structure, small in size (only 25 cm in length on the side) and light in weight, ensuring that it does not produce excess substances and gases that affect the cabin environment during operation.

The Space Application Center of the Chinese Academy of Sciences, in cooperation with the Chongqing Institute of Green and Intelligent Technology of the Chinese Academy of Sciences, has developed China's first desktop FDM printer intended for 3D printing in space (Fig. 2.5). It completed parabolic weightlessness flight tests. During the test, printing was successfully achieved in each 22 s microgravity flight environment.

2.2 Electron Beam Freeform Fabrication Technology

The working principle of Electron Beam Freeform Fabrication (EBF) technology is summarized as follows.

Fig. 2.5 Microgravity FDM 3D printer developed by Chinese Academy of Sciences

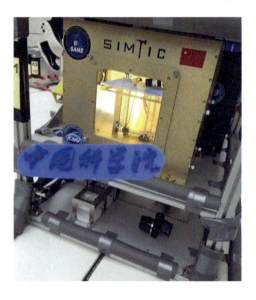

First, import the required 3D model into the printer.

Then use a high-energy electron beam as the energy source to melt the metal wire according to the set shape, forming a metal melt pool, and move along the set trajectory. The electron beam continuously scans, fusing the metal melt pools together to form a specific shape.

After that it continues to use the electron beam for melting, scanning, and deposition on the deposited metal shape until the complete 3D model is finally formed (Fig. 2.6).

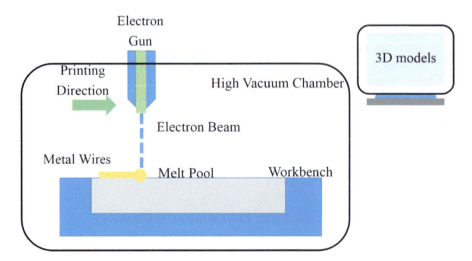

Fig. 2.6 Working principle of Electron Beam Freeform Fabrication (EBF) technology

There are two necessary conditions for Electron Beam Freeform Fabrication 3D printing technology: one is to provide a detailed 3D model of the object to be created. The other is that the material used to print the created object must be compatible with the electron beam, that is, it can be melted by the electron beam.

The raw material used in Electron Beam Freeform Fabrication 3D printing technology is metal. For the aerospace industry, titanium alloy with excellent high-temperature resistance and high strength is often used as the raw material for Electron Beam Freeform Fabrication 3D printing technology. The advantages of Electron Beam Freeform Fabrication technology mainly lie in its fast printing speed (the electron beam has a deeper penetration depth), not limited by the external shape of the 3D model and the complex internal cavity, and can form specific mixed complex structure products; no additional support structure is needed, and no post-processing for support structure cleaning is required; the equipment works in a high vacuum environment, unaffected by foreign impurities; the temperature is constant during the printing process, the shape stability is good, the residual stress is low, and the mechanical properties are far superior to ordinary cast parts, etc.

Its main disadvantages are poor forming accuracy and surface quality, and it is not suitable for materials that cannot be melted by an electron beam. Therefore, Electron Beam Freeform Fabrication technology is mainly used to prepare structural parts and support parts that do not require high shape accuracy and require high strength for the printed object.

In space applications, it has the advantage of meeting the high vacuum working conditions of electron beam wire deposition molding technology. At the same time, although limited by conditions such as microgravity, the function of the electron gun is basically unaffected and can be used normally. However, when using metal powder under the microgravity conditions in space, the metal powder may float in a free state, making it difficult to accumulate and position, and it is easy to splash around when the electron beam is incident, causing safety hazards. It is not feasible for the application of electron beam wire deposition molding technology using metal powder in space. Therefore, replacing metal powder with metal wire is a good spatial solution. The electron beam wire deposition molding technology using metal wire material is a more ideal space 3D printing technology scheme (Fig. 2.7).

In recent years, the Langley Research Center of the National Aeronautics and Space Administration (NASA) has carried out the development of microgravity electron beam wire deposition molding 3D printing technology and its printer.

2.3 Selective Laser Sintering Technology

The working principle of Selective Laser Sintering (SLS) 3D printing technology is: first establish a three-dimensional model of the printed object, and discretize the three-dimensional model into multiple small facets. During the printing process, a high-power laser is used to fuse metal, ceramic, polymer and other material powders into blocks with the required three-dimensional facet shape. Then, according to the

Fig. 2.7 Main functions of the electron beam wire deposition molding 3D printer and printed metal structural parts

preset model, the laser is stepped in three-dimensional space to complete the printing layer by layer. After printing a layer, the powder bed drops by one layer thickness, and the automatic roller adds a new layer of powder material on top of the previous one, repeating the above process, finally forming the designed three-dimensional object, and then removing the remaining uncured powder (Fig. 2.8). During the printing process, the temperature of the workbench is heated as a whole to slightly below the melting temperature of the material powder to reduce thermal deformation and enhance the bonding force between each layer. For metal powder, in order to protect it from being oxidized during the printing process, a certain protective gas needs to be filled in the printer. The protective gas used varies depending on the raw material.

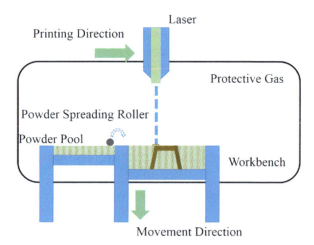

Fig. 2.8 Working principle of Selective Laser Sintering technology

Selective Laser Sintering 3D printing technology can be widely used for printing materials such as plastics, metals, ceramics, and polymers. For spacecraft applications, metals, ceramics, and polymers are the preferred SLS printing materials. Compared with other 3D printing technologies, products made by Selective Laser Sintering 3D printing technology have excellent mechanical properties, anti-aging, good environmental adaptability, etc., especially suitable for personalized production of components and parts with high mechanical structure strength and complex structure, with the outstanding advantages of no need for molds, excellent performance, and lower cost compared to mass production.

However, the defects of Selective Laser Sintering technology are also obvious, mainly reflected in the long cooling time, sensitivity to heating temperature and laser parameters, requiring a long time of process exploration. In addition, the laser system may cause additional dangers, which is not conducive to widespread use in homes or schools. The cost of 3D printers is relatively high, etc. In the aerospace industry, Selective Laser Sintering technology can be used to prepare components or parts on Earth and for use in space. When directly used in space, it will be restricted by protective gases, powder raw materials, etc., and will no longer be applicable, and it is necessary to carry out space adaptability research and verification.

2.4 Stereolithography Technology

As one of the earliest 3D printing technologies, Stereolithography (SLA) has a long history. Its working principle is as follows: first, a three-dimensional model of the design object is established and divided into layers. Then, a laser of a specific wavelength and intensity is applied to the surface of the photosensitive resin. The laser moves horizontally to solidify the photosensitive resin, forming a layer of the graphic. Then, the lifting platform moves vertically by the distance of one layer, and the laser begins to print and solidify the next layer of resin material. This process is repeated layer by layer to form the final desired 3D object (Fig. 2.9). Because that the photosensitive resin has high viscosity, a scraper is needed to evenly coat each layer after solidification, thus achieving good printing accuracy and surface roughness.

The greatest technical advantage of Stereolithography technology is its high printing accuracy (it can achieve a printing accuracy of 0.05 mm, and some printers can achieve 0.01 mm or even better printing accuracy), and good surface roughness. In addition, this technology also has advantages such as almost no waste of raw materials, fast forming speed, high printing efficiency, low printing cost, and the ability to obtain transparent or semi-transparent objects.

The main drawback of Stereolithography technology is that it requires support, otherwise the printed object may deform due to physical or chemical reactions. As the photosensitive resin is used, the mechanical properties of the printed object are poor and cannot be used for printing support structures. The requirements of storage conditions for liquid photosensitive resin are relatively strict. It has certain toxicity

Fig. 2.9 Working principle of Stereolithography (SLA) technology

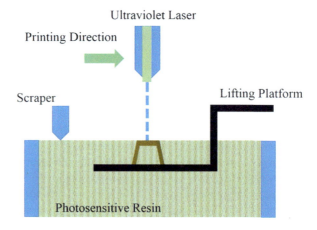

and volatility for the storage item. As a result, the impact on the environment needs to be considered.

Therefore, Stereolithography 3D printing technology is often used for spacecraft models, various component models (Fig. 2.10), especially for the model making of precision parts, with the advantages of fast forming, low cost, and easy personalized production. For some parts in the development process of spacecraft, and even the whole model of the satellite, the feasibility and assemblability can be checked through Stereolithography technology. For new designs and schemes of spacecraft, which are often complex due to space and weight requirements, after using Stereolithography technology, not only can assembly interference checks be carried out based on the 3D printed prototype model, but manufacturability discussions and evaluations can also be conducted to determine the best reasonable manufacturing process. For example, special complex parts (such as turbines, blades, impellers, etc.) can be produced in small batches through rapid investment casting, rapid sand casting, and other auxiliary technologies. Engine and other components can be tested and experimented.

Fig. 2.10 Model printed using Stereolithography technology

2.5 Concentrated Solar Power Technology

To adapt to the reality of a lack of stable and reliable energy to drive 3D printers in space, scientists have begun to explore the use of concentrated solar power 3D printing technology, which will be beneficial for the construction of lunar bases and Mars bases.

As the name suggests, concentrated solar power 3D printing technology refers to the use of collected solar energy for material melting and shaping, thereby achieving the printing of the designed model.

For industrial 3D printing on the ground, a laser is used to emit high-energy lasers to melt and sinter various raw material powders, including plastics, metals, and ceramics. Concentrated solar energy applies a similar process, applying high energy to a small area of powder, sintering it together to form a sintered powder layer of a set shape, and then repeating this process layer by layer until the printing is complete.

To achieve 3D printing with concentrated solar power, the following three steps of research need to be completed:

① Establish a physical model that can characterize material properties and printing processes, which is used to simulate the printing of lunar soil and other materials in extraterrestrial environments using concentrated solar energy.
② Design and implement a printer system, which consists of a light source, a focusing lens, a position positioning system, and a soil distributor.
③ Use modern non-destructive evaluation techniques, such as infrared imaging, X-ray analysis, and ultrasonic detection, to evaluate the printed devices, assessing the performance and effectiveness of the printed devices.

At present, scientists from various countries are conducting research on the material characteristics of sintered planetary surface soil particles in real environments, hoping to obtain their physical models. This will be a crucial step for the 3D printing construction of lunar bases or Mars bases. Concentrated solar power generation is one of the most important ways to provide clean, pollution-free, and sustainable green energy for the sintering and printing of planetary surface soil. Therefore, studying the interaction between sunlight and planetary surface matter will have a positive impact on optical mining of asteroids, thermal mining in the lunar back region, and 3D printing in-situ manufacturing technology using planetary surface matter as raw materials.

Chapter 3
3D Printing of Artificial Satellites and Rockets

3D printing technology in space not only includes 3D printing in orbit, in space stations, and in deep space exploration, but also includes the preparation of various spacecraft and their components such as satellites and rockets using 3D printing technology on Earth.

When 3D printing on Earth, it is often used to construct specific complex structures and composite materials, creating new shapes, new materials, and new devices that traditional processing techniques cannot achieve. Especially for the design and implementation of spacecraft components, 3D printing can often significantly reduce weight, reduce volume, and reduce the number of assembly parts, which has obvious advantages in reducing launch costs.

In addition, as 3D printing technology gradually moves towards application in spacecraft design and implementation, it will also bring about innovations in design technology, significant reductions in costs and expenses, and shortening of design cycles. However, it is worth noting that whether it is 3D printed satellite microwave components, antennas and other components, or 3D printed satellites themselves, it is necessary to further establish industry standards and norms for 3D printing to assess whether the material properties of printed products meet the standards for operation in space, such as resistance to impact, vibration, environmental aging, radiation, etc. It is also necessary to assess the potential impact of 3D printed materials (such as plastics and polymers) on the spacecraft environment.

3.1 Materials Suitable for Use in Space

With the development of 3D printing technology, a variety of materials suitable for 3D printing have emerged, including aluminum alloys, stainless steel, copper alloys, plastics, resins, etc. However, not all printed materials on Earth are suitable for use in space. Mainly due to the different conditions in space compared to the ground, such as cosmic environment temperature, radiation, plasma, etc. For planetary bases, in addition to temperature and radiation, it is also closely related to the humidity, atmospheric composition, dust, meteor collisions, and other environmental conditions on the planet.

At present, for spacecraft in orbit, suitable materials for 3D printing include titanium alloys, aluminum alloys, high polymer compounds, etc. With the development of 3D printing technology, there will certainly be more new types of space 3D printing materials emerging.

3.1.1 ABS Plastic

ABS plastic is a ternary copolymer composed of three monomers: acrylonitrile (A), butadiene (B), and styrene (S). The relative content of the three monomers can vary according to application needs, thus forming various resin materials. 3D printing technology using ABS plastic as raw material has been widely used on Earth, with advantages such as chemical corrosion resistance, heat resistance, good elasticity, and toughness. At the same time, ABS plastic also has advantages such as easy availability of raw materials and low cost, and is widely used in industries such as automobiles, airplanes, and ships (Fig. 3.1).

In space applications, ABS plastic is used to print various tools and daily necessities in space stations, as well as for printing and supplying various small parts (Figs. 3.2, 3.3 and 3.4). Especially for common parts or tools used inside the cabin, printing with ABS plastic has additional benefits. At this time, there is no need to consider the harsh space environment and radiation outside the cabin, but mainly consider the practicality, availability, and recyclability of the material. ABS plastic is undoubtedly a good alternative. It can prepare most daily necessities, guarantee the basic needs of astronauts and component replacement. Generally speaking, printing with ABS plastic has the characteristics of low cost, easy storage of raw materials, and easy recycling.

Fig. 3.1 3D printed complex pattern hollow plastic sphere

3.1 Materials Suitable for Use in Space

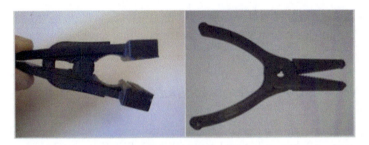

Fig. 3.2 3D printed plastic tools used in space

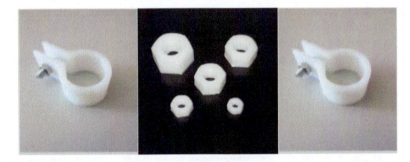

Fig. 3.3 3D printed plastic parts

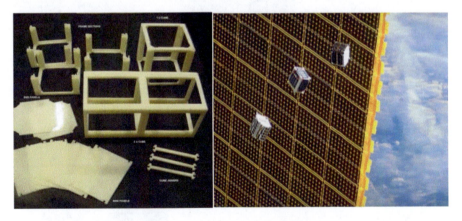

Fig. 3.4 3D printed cubic satellite shell and parts using ABS plastic

In current research, ABS plastic can also play an important role in space applications such as conductive polymer printing, electronic device and system printing, etc., by forming new high-strength thermoplastics, reinforced carbon fiber composites, and other polymer-based composites. It can even integrate sensors, antennas, and other devices, and form an integrated conformal design with the support or shell structure, creating new design concepts and application scenarios.

3.1.2 New High Polymer Compound Materials

The sample material shown in Fig. 3.5 is a high polymer compound material suitable for use in space—Polyether ketone ketone (PEKK). PEKK is a typical representative of special engineering plastics polyarylether ketone, which is a thermoplastic resin with a special structure. This is a very advanced polymer, with excellent material characteristics such as low outgassing rate, radiation resistance, high temperature resistance, chemical corrosion resistance, and solvent resistance. Existing research shows that PEKK can withstand temperature changes from -150 to $+150$ ℃; in a fire, PEKK can be used as a flame retardant, with very low smoke production rate and toxicity.

The 3D printed PEKK pipe shown in Fig. 3.6 is a material added to the printed aerospace parts, combined with the existing carbon fiber material, used for the dissipation of static charge.

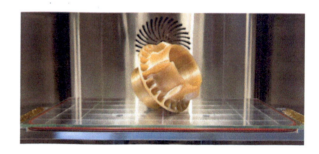

Fig. 3.5 3D printed polyether ketone ketone structure sample

Fig. 3.6 3D printed PEKK pipe tested by NASA

3.1 Materials Suitable for Use in Space

3.1.3 Metal Materials

To adapt to the special needs of vibration, temperature, etc. during the launch of spacecraft, as well as the temperature, radiation, etc. factors in space environment, various new alloy materials have been developed by aerospace agencies. Combined with the corresponding 3D printing process, it can realize the preparation of materials under different application scenarios.

The existing 3D printed special preparation alloys include aluminum-based, iron-based, copper-based, nickel-based, titanium-based and other alloys (Fig. 3.7), the main application scenarios are the preparation of spacecraft metal parts or rocket propulsion chambers and other high temperature and high pressure special application scenarios. Generally speaking, nickel-based high-temperature alloys are more suitable for these application scenarios. However, in most cases, in order to operate continuously, the working temperature of nickel-based high-temperature alloys is limited to 900 ℃ or lower. Although γ' precipitation strengthened nickel alloys can maintain certain mechanical properties at temperatures up to 1200 ℃, they may still limit the performance of the prepared parts.

GRX-810 Alloy

NASA announced a new alloy material called GRX-810 in 2023, which can withstand temperatures close to 2000 ℉ (about 1093 ℃). Its strength and oxidation resistance are twice that of the most advanced 3D printed superalloys, and its durability is more than 1000 times that.

According to NASA, GRX-810 can withstand extreme temperatures and other harsh environmental conditions. This high-temperature alloy can be used to manufacture stronger and more durable parts inside aircraft and spacecraft.

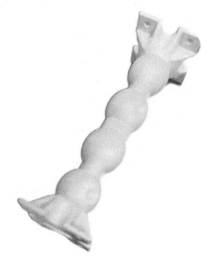

Fig. 3.7 3D printed aluminum alloy samples

The researchers said that the specific development method is to combine laser melting 3D printing technology, use laser powder bed melting to disperse nano-sized Y_2O_3 particles throughout the microstructure, and prepare a new type of oxide dispersion strengthened NiCoCr-based alloy to replace traditional mechanical processing or in-situ alloying processes.

In fact, by simply adding Y_2O_3 particles, the strength of NiCoCr has increased and the ductility has doubled. This highlights the strengthening effect provided by these oxides at high temperatures. These results show how the future alloy development combining dispersion strengthening and additive manufacturing processes will accelerate the discovery of revolutionary materials.

Manufacturer 3D Systems took this opportunity to test it on its metal 3D printing platform, confirming the excellent performance of this new material. This material will eventually appear in more metal 3D printers, producing parts that can operate normally in a wider range of environments.

High Conductivity GRCop-42 and GRCop-84

For rocket engines, due to the influence of propellants, oxidation may cause scalding, and hydrogen may cause hydrogen environment embrittlement. Thus, it is necessary to develop new alloy materials to meet these special requirements.

It is a combination of copper, chromium, and niobium for GRCop. This material is specifically optimized for high strength, high thermal conductivity, and high creep resistance, allowing for greater stress and strain in high-temperature applications. It has good low-cycle fatigue performance and prevents material failure at high temperatures. The advantages of GRCop alloy are high electrical conductivity, good strength at high temperatures, and stability during operation. Due to the high content of Cr_2Nb, GRCop-84 exhibits high strength at various temperatures. Its low-cycle fatigue performance is slightly improved.

GRCop alloys are dispersion-strengthened by Cr_2Nb precipitates produced during the powder atomization process and refined during the laser powder bed fusion metal 3D printing process. Compared with most low-alloy copper-based alloys, GRCop alloys also have improved oxidation resistance and thermal whiteness. GRCop alloys allow for hot wall temperatures $\geqslant 700$ ℃, depending on strength, creep, and low-cycle fatigue test requirements.

By using the directed energy deposition 3D printing process, bimetallic materials can be deposited on the back end of the GRCop-42 copper chamber, forming a rocket thrust chamber nozzle with a bimetallic axial joint, and achieving continuous cooling. This solves some design challenges and bolt connection design interface issues, and then the entire thrust chamber assembly is wrapped with a carbon fiber polymer-based composite material (Fig. 3.8).

In March 2023, the Terran 1 rocket developed by Relativity Space was launched from Cape Canaveral, Florida. The Terran 1 rocket has nine additive manufacturing engine combustion chambers made of GRCop copper alloy, with temperatures approaching 6000 ℉.

3.1 Materials Suitable for Use in Space

Fig. 3.8 3D printed copper alloy material GRCop-42 sample. *Source* NASA

Extreme Temperature Refractory Alloys

Generally speaking, metals with extremely high melting points are called refractory alloys. Refractory alloys can maintain their shape and overall use in a range of extreme environments, such as high acidity or alkalinity, corrosive chemicals, or extreme temperatures or pressures.

For space applications, traditional refractory metal manufacturing is usually very expensive due to high material costs, specialized powder production methods, and unique processing and connection methods. The specific difficulties in forming parts, such as high ductile-to-brittle transition temperatures, ultra-high-temperature heat treatment, specialized oxidation coatings, and non-destructive evaluation requirements, also pose significant challenges to the manufacture of refractory components.

Aerospace refractory metal parts are often thin-walled, resulting in 95%–98% of the billet being machined away. Therefore, the raw material cost of the part is 5% of the actual part, and 95% is machining waste. In addition to the raw materials, machining and waste disposal also add additional costs.

Other manufacturing methods are based on deposition, such as vacuum plasma spraying, and electro-deposition, which are often slow and expensive processes. They usually require the removal of the mandrel after deposition, which limits the complexity of the parts. Due to the difficulties of traditional refractory metal manufacturing, the number of suppliers with the necessary equipment and experience is limited.

The development of 3D printed refractory materials is a rapidly developing field, mainly using EB-PBF electron beam melting 3D printing, L-PBF laser melting 3D printing, binder jetting, EW-DED electron beam wire directed energy deposition, and LP-DED laser powder bed fusion 3D printing process. In most cases, these methods can significantly save costs and schedule. Research has found that

Fig. 3.9 Iron-nickel alloy

even considering the cost of powder raw materials, printing time, heat treatment, final processing, and waste disposal, the cost of refractory alloy parts prepared by 3D printing is significantly lower than the same parts produced by traditional manufacturing processes.

Iron-nickel alloy is a material with great potential for 3D printing in space in the future (Fig. 3.9). The range of nickel content in iron-nickel alloys is wide, and other elements may be added according to different application scenarios to enhance certain material characteristics. For example, for the 3D printing of rocket thrust chamber structures, the material needs to have good thermal conductivity, yield strength, and elongation. Especially when the cavity is filled with liquid and gaseous hydrogen propellant, the requirements for heat and structure are more complex and the challenges are great.

Stainless steel is also one of the commonly used metal materials for 3D printing. Stainless steel usually has excellent characteristics of rust resistance and corrosion resistance. For 3D printed stainless steel materials for space applications, the requirements are more stringent, often requiring high pressure resistance, hydrogen embrittlement resistance, oxidation resistance, etc. Different metal elements can be added to improve corresponding characteristics. Austenitic stainless steels, such as 304, 310, and 316 stainless steels, have good hydrogen embrittlement resistance, but their yield strength is low.

Invar steel is a type of iron-nickel alloy (Fig. 3.10), which deforms very little with temperature changes and is particularly suitable for components with strict temperature deformation requirements. Also due to its excellent temperature characteristics, it is an important alternative material in the preparation process of spacecraft components where temperature changes are large but high precision is required.

According to the current state of technological development, new alloys that can be printed include GRCop-42, GRCop-84, NASA HR-1, GRX-810, C-103, C-103 CDS, Mo, and W, mainly used for 3D printing spacecraft components to achieve higher performance.

Fig. 3.10 Invar steel

For further technological development, it is necessary to conduct targeted research according to specific application scenarios in space, in order to obtain alloy materials with unique excellent characteristics, while taking into account the process characteristics of 3D printing.

3.1.4 Piezoelectric Materials

Piezoelectric materials generally refer to crystal materials that will generate voltage between the two ends of the material when subjected to pressure, widely used in the preparation of sensors and other electronic devices. 3D printed piezoelectric materials can be used for the preparation of electronic components and photonic components in space, used to form the required electronic or photonic circuits and functional modules.

The base materials for 3D printing piezoelectric materials include plastics, glass, silicon wafers, polymers, etc. Carbon nanotubes, silver, gold, titanium dioxide, silicon dioxide, and various 3D printing inks containing surfactants that are stable at work are gradually emerging (Fig. 3.11).

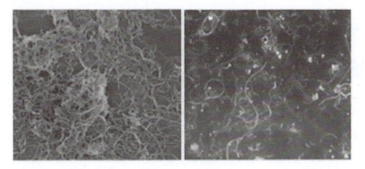

Fig. 3.11 3D printed carbon nanotube droplet ink point

To achieve the printing and preparation of electronic components in space, it is first necessary to select and develop printable, stable performance raw materials, then prepare printing equipment and printing processes that are compatible with them, and complete the printing, testing, and verification of circuit modules.

3.2 3D Printed Satellite Components

Satellites, short for man-made earth-orbit satellites, are currently the most widely used and applied spacecraft. Satellites operate in different Earth orbits, carrying out various space missions, and are an important means for mankind to step into space and explore the vast universe.

According to the functions and uses of satellites, they can be divided into communication satellites, weather satellites, ocean satellites, navigation satellites, observation satellites, reconnaissance satellites, etc.

According to the different heights of the satellite orbits, they can be divided into low-orbit satellites (polar satellites, orbit height less than 5000 km), medium-orbit satellites (orbit height from 5000 to 20,000 km), and high-orbit satellites (geosynchronous orbit satellites, orbit height above 20,000 km).

According to the different weights of satellites, they can be divided into large satellites (more than 3000 kg), medium satellites (1000–3000 kg), small satellites (100–1000 kg), and microsatellites (less than 100 kg), etc. (Fig. 3.12).

Fig. 3.12 Different types of satellites. *Source* China Aerospace Science and Technology Corporation

3.2 3D Printed Satellite Components

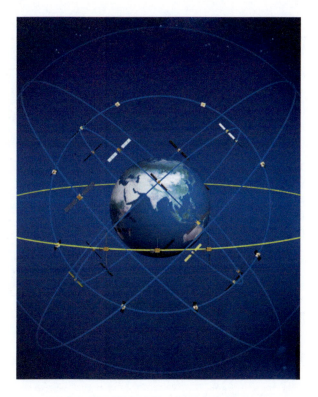

Fig. 3.13 Beidou series navigation satellites

The first artificial satellite launched by China—Dongfanghong-1 satellite has a mass of 173 kg. The Beidou series of satellites are networked by multiple satellites operating in geosynchronous stationary orbits, inclined geosynchronous orbits, and medium earth orbits, providing all-weather, all-time, high-precision positioning, navigation, and timing services for global users (Fig. 3.13).

When using 3D printing to design and implement satellite parts, the main considerations are the improvement of part performance and the implementation of new designs.

The performance improvements that 3D printing can bring include integrated design, lightweight design, compact structure design, etc. (Figs. 3.14 and 3.15), resulting in significant reductions in product weight and volume or significant improvements in mechanical and thermal performance.

The implementation of new designs mainly depends on the 3D printing process's ability to implement complex structures, especially the implementation of special stacked structures inside the cavity, which can achieve structures that traditional mechanical processing techniques cannot achieve, thereby innovating design ideas and concepts.

Before using 3D printing technology to design satellite parts, it is first necessary to study the characteristics of materials and equipment and obtain relevant information to meet performance requirements. For this, it is necessary not only to determine the

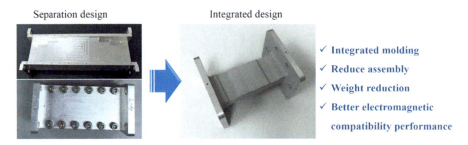

Fig. 3.14 Implementing integrated design using 3D printing

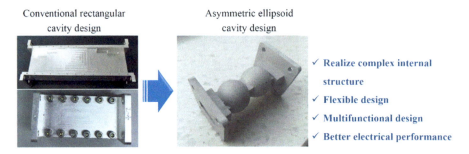

Fig. 3.15 Implementing new structure design using 3D printing

appropriate process parameters of the printing equipment, but also to conduct material tests. For example, for applications that require certain mechanical properties such as structural strength or material properties such as density, the maximum size of the printed sample to the maximum force it can withstand before breaking is essential information needed before starting the design.

After fully testing the capabilities of the printing equipment, material parameters, and structural parameters, a database of materials and equipment can be established for subsequent design.

In the process of experimentation, the parameter database can be iteratively updated according to the characteristics of the product to meet the design and implementation of parts with more complex and higher performance requirements.

3.2.1 3D Printed Filters

Filters are a common type of microwave component in satellites payload system, whose main function is to filter certain frequency RF signals and are widely used in communication satellites, remote sensing satellites, and others.

What unique advantages can the combination of 3D printing and filter design produce?

3.2 3D Printed Satellite Components

Firstly, Integrated design can be achieved through 3D printing, eliminating the cumbersome steps of traditional mechanical processing technology that requires the filter to be dissected, processed internally, and then assembled. This reduces assembly costs and time, and most importantly, results in weight reduction, which is an extremely important design factor for satellite applications. Furthermore, through the integrated design of filter devices, the assembly details such as waveguide flanges can be further reduced. This is significant for millimeter wave and even higher frequency microwave systems in terms of weight reduction and improvement of device performance discontinuity. Finally, 3D printing technology can be used to design complex and irregular structures beyond conventional shapes, to enhance performance and greatly increase the flexibility of engineers' designs.

Figure 3.16(a) is an irregular filter made of stainless steel using 3D printing. Compared to traditional coaxial and rectangular cavity designs, this ellipsoidal structure has the advantages of higher Q values and smaller volume. At the same time, the metal wall thickness is only 2 mm, greatly reducing the weight of the cavity processed by traditional mechanical methods and reducing the assembly process. Figure 3.16(b) is a 3D printed irregular structure duplexer.

However, it should be noted that currently, due to the limitations of 3D printing precision and surface roughness, insufficient precision and poor surface roughness are the main factors limiting its application in filters, and other microwave components or higher frequency systems. With the advancement of technology in the future, 3D printing will be more widely and commonly used in satellite components and systems.

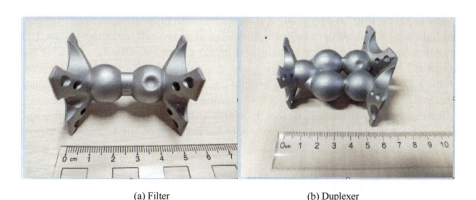

(a) Filter (b) Duplexer

Fig. 3.16 3D printed irregular structure filter and duplexer

3.2.2 3D Printed Antennas

Antennas are one of the most essential components of satellites and other spacecraft. For communication, navigation, data transmission and other applications in space industry, antennas play an important role of transmitting and receiving electromagnetic wave signals. They are the "ears" and "mouth" of spacecraft.

The strength of the signal that an antenna could receive is directly related to the size of the antenna. Generally speaking, the larger the physical size of the antenna, the weaker the signal it can receive, that is, the higher the receiving sensitivity. Therefore, for spacecraft, the design and implementation of antennas is of paramount importance. Whether a large antenna can successfully deploy in orbit determines whether the spacecraft has "good hearing".

Can antennas be 3D printed? The answer is yes. Using 3D printing technology, we can design a variety of new forms of antennas (Fig. 3.17).

For space applications, 3D printing technology is equally important for the design and preparation of spacecraft antennas.

Until now, we are no longer limited by traditional mechanical processing, and can use 3D printing technology to print a variety of antennas with new forms, structures, and materials, achieving performance improvements.

The European Space Agency and SWISS to 12 of Switzerland have used 3D printing technology to achieve a space dual-reflector antenna, improving the processing precision of the antenna, reducing costs, and shortening the development and delivery time. At the same time, the use of 3D printing technology has also increased the flexibility of antenna design (Fig. 3.18).

Using 3D printing to design support structures with high strength, small volume, and light weight can achieve lightweight design of spacecraft antennas. Figure 3.19 is Thales Alenia Space of France using 3D printing to achieve the Koreasat 5A and Koreasat 7. The antenna support structure of remote communication satellites adopts a hollow integrated design, reducing weight by 22%, saving nearly 30% in cost, and shortening the production cycle by 1 to 2 months.

Researchers from Second Research Institute of China Aerospace Science and Industry Corporation Limited used photopoly merization molding technology to print

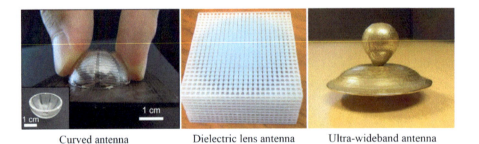

Curved antenna   Dielectric lens antenna   Ultra-wideband antenna

Fig. 3.17 Various 3D printed antennas

3.2 3D Printed Satellite Components

Fig. 3.18 3D printed dual-reflector antenna

Fig. 3.19 3D printed antenna support structure for remote communication satellites

a broadband dielectric resonant antenna for navigation satellites (Fig. 3.20), providing a new approach to satellite antenna design.

Fig. 3.20 3D printed broadband dielectric resonant antenna

3.2.3 3D Printed Loop Heat Pipe

A loop heat pipe is a closed-loop heat pipe, generally composed of an evaporator, a condenser, a liquid storage device, and vapor and liquid pipelines. It is widely used in energy distribution, usage, and residual heat recovery. The planar loop heat pipe is highly efficient in heat transfer. But its cylindrical conventional design makes it difficult to install on electronic equipment boxes where heat pipe networks are usually distributed.

Because the loop heat pipe has many components, using 3D printing will help reduce the cost of building this technology. At the same time, realizing the planar design of the loop heat pipe based on 3D printing technology will also make the equipment easier to install.

The design in Fig. 3.21 uses a "sandwich" structure with the loop heat pipe located between the electronic equipment boxes. The biggest technical challenge lies in the implementation of the inner core structure of the heat pipe. The inner core is located in the center of the heat pipe, and the hole width in its porous design should be 1–2 μm. However, the machining accuracy of 3D printing technology can generally only reach 20 μm. Using holes larger than 20 μm means that the pressure inside the heat pipe will drop to a lower level, unable to maintain the capillary force required for loop operation. Researchers are working hard to overcome technical difficulties, and the European Space Agency has achieved 3D printing technology with a hole size of 10 μm.

The experimental equipment made of stainless steel completed environmental testing. In future research, the hole size will be further reduced, and weight reduction design will be carried out.

This 3D printed loop heat pipe has a more compact structure, suitable for thermal control design of small satellites and cube satellites, and can reduce cost and design complexity.

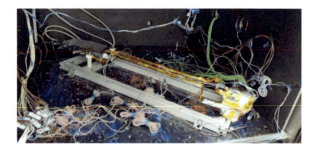

Fig. 3.21 3D printed satellite loop heat pipe. *Source* European Space Agency

3.2.4 3D Printed Atmospheric Spectrometer Mirror

Considering the new manufacturing possibilities provided by 3D printing technology, redesigning existing high-performance components is a possible way to evaluate it. For example, the European Space Agency collaborated with the Dutch TNO Research Institute to redesign optical mirrors using 3D printing. This device is a key component of the atmospheric spectrometer of the Tropospheric Monitoring Instrument (Tropomi), which was put into use in the European Space Agency's Sentinel-5P mission in 2016.

The Tropomi developed by the Netherlands identifies trace gases (gases with concentrations less than 10^{-6} in the atmosphere) by filtering out specific spectral lines from light passing through the Earth's atmosphere. To do this, a pair of high-precision deformable mirrors are needed to form an "optical cavity", focusing the incident light onto a slit in the instrument, thereby forming a spectroscopic grating. Since the optical quality of the mirror also needs to be accurately achieved, this process sets strict goals for 3D printing.

The original design of the reflector was composed of 284.6 g of aluminum and a layer of nickel-phosphorus coating. During the 3D printing process, the reflector was redesigned using selective laser melting technology, using titanium alloy as the manufacturing material. The final design of the reflector is streamlined, with a weight reduction design that removes unnecessary weight, resulting in a final weight of only 127.7 g, using the same coating (Fig. 3.22). Most importantly, the optical performance of the reflector did not decrease during this process. This work provides new ideas and implementation methods for the weight reduction design of components and systems working in space in the future.

Original design	New design	Selective laser melting
Weight: 284.6g	Weight: 129.7g	Weight: 127.7g
Material: Aluminum	Material: Titanium	Material: Titanium alloy (Ti_6Al_4V)

Fig. 3.22 3D printed starboard reflector. *Source* European Space Agency

3.2.5 New Welding Technology Based on 3D Printing

When spacecraft operate in space, the connections between different materials, different devices, and between components and cavities often use welding, which brings a big problem. The bonding structures are affected by the drastic changes in temperature in space for the rate of thermal expansion and contraction between different materials is different. It can lead to fractures or damage between materials and devices. Although during the development process on Earth, aerospace engineers will conduct various harsh environmental tests on the spacecraft to ensure its safe and reliable operation in space. However, considering the hundreds or even ten thousands of connection structures on the spacecraft, it still poses a great risk of failure to the spacecraft in orbit.

Which method can reduce the tiny connection parts and joints, and improve the reliability of welding? The application of 3D printing technology in new welding technology is promising.

Take the heat exchanger as an example to illustrate the advantages of the new welding technology based on 3D printing. Temperature is always a particularly noteworthy issue in space. Affected by solar radiation, the extreme temperature that the spacecraft faces may change by hundreds of degrees Celsius. The heat exchanger is a particularly useful part on the spacecraft, mainly used to remove excess heat, or to absorb more heat to help the spacecraft maintain a stable internal temperature. The traditional heat exchanger design often includes long curved pipes, connected to metal plates with brackets and epoxy resin. These pipes contain many interconnected structures, introducing many potential failure points.

Based on ultrasonic 3D printing technology, the integrated preparation of heat exchangers can be realized (Fig. 3.23). The new design eliminates dozens of small parts and joints, reducing the probability of failure in the extreme temperature environment of space.

How to achieve welding between metals in the heat exchanger? This requires reliance on ultrasonic technology, using ultrasonics and friction to form solid-state adhesion between metal and metal (Fig. 3.24). First, the metal layer is laminated

Fig. 3.23 3D printed metal heat exchanger

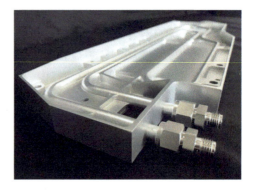

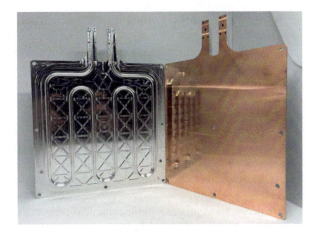

Fig. 3.24 Solid-state adhesion between metal layers

on another metal part (such as the base plate), and the friction between the metals is caused by constant pressure and ultrasonic vibration. Then the temperature is increased to remove the oxide on the metal surface, allowing pure metal to directly contact pure metal. Finally, solid-state atomic bonds are formed, welding the metals together. At this time, the temperature at which the metal combines with the metal is much lower than the melting temperature of the metal, the required heat is greatly reduced, thereby shortening the welding and manufacturing cycle.

This technology is very useful for the advanced manufacturing of spacecraft. It can formulate different printing and welding schemes according to the properties of different metals. It will not cause the liquefaction and mixing of metals in this process. Furthermore, this technology makes it possible to embed the best sensors between metals, materials, and devices. During the embedding process, the sensors will not be damaged by high-temperature welding. Accurate real-time data information of the materials can be obtained.

3.3 3D Printing of Satellite Structure

Compared to 3D printing of satellite parts and functional parts, it is relatively easier to construct the overall structure of a satellite using 3D printing technology. This is because, in addition to mechanical properties, functional parts often have specific requirements for printing size accuracy, surface roughness, and even electrical conductivity. Using 3D printing for structural part design and implementation, in addition to reducing weight, can also improve structural mechanical performance through integrated design.

Using 3D printing technology to prepare the main structure of the satellite is one of the main development directions at present. Complex structural designs that cannot be achieved with traditional processes can reduce weight and the number of structural parts while ensuring the mechanical performance of the main satellite structure.

Space lattice structures are often used to characterize and describe the microscopic atomic/ion structure of material crystals. For example, many metals have a body-centered cubic space atomic lattice structure. Figure 3.25 shows a typical crystal unit cell space lattice structure, which is the atomic space arrangement structure of diamonds and other crystals.

In 3D printing technology, a structure similar to the crystal lattice structure is used as the basic unit of the three-dimensional space structure to achieve lightweight design while ensuring good mechanical performance. Then, through the periodic arrangement of the lattice structure, a macro three-dimensional structure design is realized (Fig. 3.26).

The preparation process and mechanical properties of each lattice structure need to be studied theoretically and experimentally before use to ensure that the printed structure has the expected performance.

The 3D printed space lattice structure has been applied to the implementation of the main structure of the satellite. Figure 3.27 shows a lattice structure designed by the General Department of the China Academy of Space Technology (the Fifth

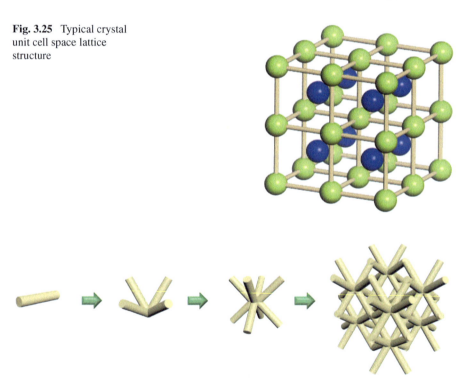

Fig. 3.25 Typical crystal unit cell space lattice structure

Fig. 3.26 Formation of spatial lattice structure

3.3 3D Printing of Satellite Structure

Fig. 3.27 Metal lattice structure. *Source* Xi'an Platinum Special Additive Technology Co., Ltd.

Academy of Aerospace), which is a special three-dimensional structure of metal 3D printing, realizing lightweight design to replace traditional solid plates.

Figure 3.28 shows the main structure of a satellite composed of lattice structures as basic units. When using 3D printed lattice structures for periodic expansion, it has technical advantages such as large size and high local accuracy. Using aluminum alloy 3D printing technology, the General Department of the China Academy of Space Technology has reduced the weight of the main structure of the satellite, and the number of parts of the entire satellite structure has been reduced to 5, and the design and preparation cycle has been shortened to 1 month. The overall structure has large spatial dimensions and thin skins. The interior of the "sandwich" structure formed by the skin is densely distributed with a spatial grid structure. According to the force situation, there are dense and sparse areas in this grid structure. The grid density is denser at the installation hole position, and sparser at other positions. While achieving part weight reduction, the strength of the parts is ensured. The diameter of the grid structure is small, most of the overall diameter is 0.5 mm, and the diameter of the local dense area is 0.25 mm.

Fig. 3.28 3D printed skin structure formed by lattice structure. *Source* Xi'an Platinum Special Additive Technology Co., Ltd.

The lattice brackets shown in Figs. 3.29 and 3.30 are also an important part of the load-bearing structure on satellites. Through preliminary structural research and optimized design, this lattice bracket has significantly improved its lightweight performance while ensuring its mechanical properties.

Currently, the Qiancheng satellite, which uses a 3D printed lattice structure as the main structure of the satellite, was launched into orbit in August 2019 by the Jielong Yaoyi rocket of China Aerospace Science and Technology Corporation at the Jiuquan Satellite Launch Center. The on-orbit operation is stable, verifying the practical application of the 3D printed lattice structure in the main load-bearing structure of the satellite.

The hydrazine bottle bracket, as a large-scale load-bearing part based on 3D printed lattice materials, is the supporting structure of the hydrazine bottle in the on-orbit propulsion system resource platform of the remote sensing satellite, and is also the key load-bearing part of the entire thrust system. The 3D printed remote sensing satellite hydrazine bottle bracket shown in Fig. 3.31 is assembled into a structure with a diameter of 1200 mm by printing 6 pieces separately. The 3D printing process has completed the printing of the skin and large-size lattice structure parts, with a

Fig. 3.29 3D printed lattice bracket formed by lattice structure (Ⅰ). *Source* Xi'an Platinum Special Additive Technology Co., Ltd.

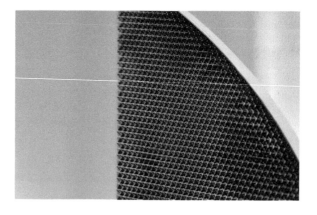

Fig. 3.30 3D printed lattice bracket formed by lattice structure (Ⅱ). *Source* Xi'an Platinum Special Additive Technology Co., Ltd.

3.4 3D Printed Rockets

Fig. 3.31 3D printed remote sensing satellite hydrazine bottle bracket formed by lattice structure. *Source* Xi'an Platinum Special Additive Technology Co., Ltd.

minimum wall thickness of 0.5 mm and a lattice wire diameter of 0.5 mm; compared with traditional processes, the overall weight of this structure has been reduced by more than 60%.

3.4 3D Printed Rockets

A rocket is a flying vehicle that propels forward by generating a reaction force through the ejection of a medium. Strictly speaking, rockets are mainly used to carry spacecraft into space and are the means of transportation for space flight, and are not classified as spacecraft themselves.

Rockets can be divided into launch rockets, sounding rockets, rocket missiles, rocket artillery, etc., according to their uses. This book mainly discusses launch rockets, which are used to carry satellites, space stations, space rovers, and other spacecraft into predetermined orbits in space.

Rockets can be divided into chemical rockets, electric rockets, nuclear rockets, etc., according to their power sources. Currently, liquid propellant rockets (simply called liquid rockets) in chemical rockets are widely used in the aerospace industry.

At the beginning of the twentieth century, Russian scientist Konstantin Tsiolkovsky theoretically proved that multi-stage rockets could successfully overcome Earth's gravity and enter space. American rocket expert Robert H. Goddard further combined space theory and rocket technology to calculate that a rocket must have a speed of 7.9 km/s to overcome Earth's gravity and enter space, thereby establishing a theoretical basis for space launch rockets.

The working principle of a rocket can be briefly described as burning liquid propellant in the engine combustion chamber, generating a large amount of high-pressure gas, and ejecting it at high speed from the engine nozzle, thereby generating a reverse thrust and propelling the rocket forward at high speed. Of course, the actual rocket control system is quite complex. However, understanding the basic working principle can help us understand the role and possible applications of 3D printing technology in it.

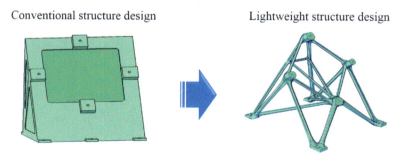

Fig. 3.32 Structural optimization design

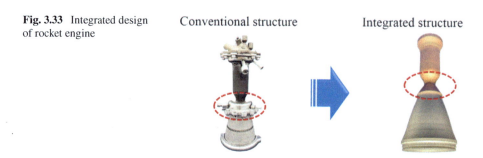

Fig. 3.33 Integrated design of rocket engine

3D printing has broad application prospects in rocket development. 3D printing can be used for the weight reduction design of rockets. By printing on the rocket shell and engine casing to optimize the structure, lightweight design can be achieved (Fig. 3.32). 3D printing can also be used to improve the mechanical and thermal performance of rocket engine components. For example, using an integrated structure to increase the mechanical strength of the engine structure, and using new materials printed to enhance thermal conductivity, etc. (Fig. 3.33). For conventional separated structures, they are often processed separately and then connected together by welding or using flanges and screws, which can lead to risks of cracking and insufficient mechanical strength under extreme conditions. Integrated design can not only reduce weight, but also has the additional benefit of enhancing mechanical strength.

3.4.1 3D Printing Rocket Engine Components

Future lunar landers may be equipped with 3D printed rocket engine parts, which can help reduce overall manufacturing costs and shorten production time. Similarly, developing lighter and more efficient liquid rocket engine components is of great significance for launch missions to the moon, Mars, and further planets.

Based on this, aerospace agencies and research organizations in various countries have begun to carry out preliminary technical verification of 3D printing technology

3.4 3D Printed Rockets

for rocket components in recent years, such as 3D printing and testing of rocket engine components, and will gradually promote its application in the future.

The rocket thrust chamber is the component in the rocket engine that completes the conversion of propellant energy and generates thrust. Relatively speaking, the structure of the solid rocket thrust chamber is simpler. The liquid rocket thrust chamber mainly consists of an injector, a combustion chamber, and a nozzle. For example, the Saturn V rocket launched by the United States, the engine of the first stage rocket is filled with 2000 t of kerosene and liquid oxygen, and its working time is 150 s. All the fuel will be pumped from the fuel tank to the rocket engine nozzle within 150 s, outputting a thrust of 33,000 kN, almost equivalent to the total thrust of 91 J-20 aircraft using WS-15 engines.

The rocket thrust chamber is the most expensive component of the liquid rocket engine, not only complex in structure and long in manufacturing time, but also high in performance requirements. The rocket thrust chamber is also the heaviest component in the rocket engine, and its weight increase will significantly increase the cost of the launch mission. If the weight of the rocket thrust chamber can be reduced while maintaining performance, it will be of great benefit to cost reduction and efficiency improvement.

Figure 3.34 shows the 3D printing technology and structural composition used for the rocket thrust chamber components, which are composed of a combustion chamber, a nozzle, and a joint. In the rocket engine, the fuel and oxidizer mix and burn in the combustion chamber. This combustion produces hot exhaust, which accelerates the airflow through the nozzle and generates thrust. The main goals in the process of realizing 3D printed rocket thrust chambers are: ① Develop the application of Directed Energy Deposition (DED) technology in the preparation of complex large nozzles; ② Develop composite material coatings that can be used to reduce weight and provide structural strength for the thrust chamber; ③ Develop dual-metal, multi-metal 3D printing technology.

To reduce the cost of rocket thrust chambers, composite materials were used to replace some traditional metal structural parts on the thrust chamber. Advanced

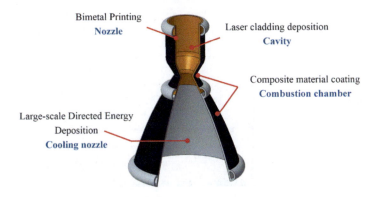

Fig. 3.34 Structure of 3D printed rocket thrust chamber components

3D printing technology was used to print combustion chambers and nozzles, and innovative mechanical methods were used to combine the two, rather than completing the connection based on traditional metal joints (Figs. 3.35 and 3.36). The surface of the 3D printed copper combustion chamber has a composite material coating. The composite material coating is made of carbon fiber, which can provide structural support for the combustion chamber. Compared with traditional metal protective sleeves, this carbon fiber-made coating achieves a weight reduction of up to 50%. The nozzle is printed using Directed Energy Deposition (DED) technology. DED technology first uses a multi-axis mechanical arm to deposit material on the surface. Then uses a laser to melt, deposit and solidify the raw material, forming the designed structure. This printing technology significantly reduces the manufacturing cycle of rocket engine nozzles, and compared to traditional processes. The preparation time was reduced from nearly two years to several months. Using 3D printing technology can also greatly reduce the number of parts (especially the number of connectors).

Figures 3.37, 3.38 and 3.39 show the shapes and structural components of 3D printed rocket thrust chamber components at different stages of NASA's technology development. As can be seen, in early research, small, separate device printing was first carried out, still with a large number of connection structures and equipment. To ensure a large thrust force, stringent design requirements were put forward for the connection structure. Thus it has a complex structure and a large weight. Subsequently, as the thermal load and structural load became more challenging, the thrust chamber developed towards a larger scale and higher cavity pressure (Fig. 3.39).

Fig. 3.35 Two-segment cavity structure printed using laser cladding technology

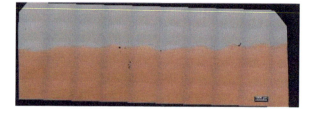

Fig. 3.36 Appearance of mixed metal printing connection

3.4 3D Printed Rockets   47

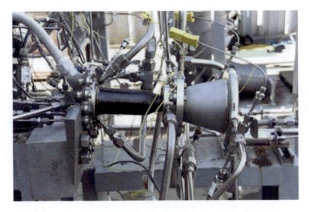

Fig. 3.37 High-strength iron-nickel high-temperature alloy nozzle with composite material coating printed using directed energy deposition technology. *Source* NASA

Compared with the structure using bolt connections, the integrated design using 3D printing obviously has the advantages of lighter weight and better structural strength.

Figure 3.40 shows the printing process of a 177,929 N thrust chamber structural component. The structural component is printed with a copper alloy, the main material is GRCop-42, which has excellent electrical conductivity. The thickness of the structural component cavity wall is determined by the designed thrust level.

In May 2021, NASA reported that its 3D printed rocket engine hardware passed cold spray and hot combustion tests (Fig. 3.41). This hardware is mainly used for lightweight design of thrust chamber components, and is printed from a mixture of two metals. In the test, the hardware was started 8 times in total, with a total hot start duration of 365.4 s. In all the tests conducted, the pressure endured by the main

Fig. 3.38 High-temperature combustion test of 3D printed rocket thrust chamber components. *Source* NASA

Fig. 3.39 Different thrust chamber structures using 3D printing technology

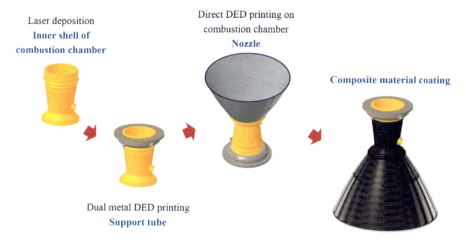

Fig. 3.40 Printing process of a 177,929 N thrust chamber structural component

combustion chamber was as high as 5170 Pa, and the calculated hot gas temperature was close to 6200 ℉ (3427 ℃). Three different carbon composite nozzles designed for 31,137 N of thrust were also tested, and it was proven that they could withstand extreme environmental conditions, with measured nozzle temperatures exceeding 4000 ℉ (2204 ℃). The successful testing of this 3D printed rocket engine hardware demonstrates the capabilities of new manufacturing technologies and their future applications. This 3D printing technology plays a significant role in reducing hardware manufacturing time and costs. This test is also a key step in reserving hardware for future lunar and Mars missions.

3.4 3D Printed Rockets

Fig. 3.41 3D printed rocket engine components and their testing process

3.4.2 Existing 3D Printed Rockets

With the support of a German data analysis company, a team in the UK has completed the manufacture of a simple rocket using 3D printing technology (Fig. 3.42). The rocket weighs 3 kg, is as tall as an average adult, and costs about 6000 pounds. The team stated that after receiving new investment, they would consider launching this 3D printed rocket into the sky. After the launch is completed, the rocket's autopilot system will guide the rocket back to Earth, and the camera inside will record the entire process.

As shown in Fig. 3.43, with the support of the UK Space Agency, the UK space launch company Orbex used selective laser melting 3D printing technology to complete the printing of the engine part of a large rocket. Titanium and aluminum alloys were used as the main raw materials to withstand the extreme temperatures and pressures required during the rocket launch process. This rocket engine is currently the largest 3D printed rocket engine announced in the world. The entire engine is 3D printed, expected to withstand extreme temperature and pressure changes, and the overall weight is reduced by 30%. In June 2022, the company announced that it had completed the full integration of the rocket and set it up on its platform, and will conduct extensive testing, launch drills, and adjustments to the launch program.

Fig. 3.42 3D printed rocket

Fig. 3.43 3D printed large rocket engine

3.4.3 3D Printed Rocket Launch and Landing Devices

In March 2021, at the Swift Camp in Bastrop, Texas, USA, the "Artemis" team, composed of students from various American universities, demonstrated the performance of their designed and completed 3D printed rocket launch and landing pad. This rocket launch and landing pad is mainly used to support the hot rocket engine and is expected to solve a series of problems caused by lunar dust during the rocket launch and landing process on the lunar surface. The "Artemis" team named it the "Lunar Plume Alleviation Device" (Figs. 3.44 and 3.45).

The team received support from a student collaboration program established by NASA's Marshall Space Flight Center and Arizona State University. With the help of NASA's "From Moon to Mars" Planetary Autonomous Construction Technology (MMPACT) project, the American ICON company, and the Texas A&M University

Fig. 3.44 3D printed Lunar Plume Alleviation Device

3.5 3D Printed Engine Components

Fig. 3.45 3D printed Lunar Plume Alleviation Device test

sounding rocket team, the first generation of "Artemis" members obtained funding to print and test small prototype machines.

3.5 3D Printed Engine Components

In the manufacture of rocket engines, the use of 3D printing technology can enhance mechanical performance and heat dissipation performance, and has a huge advantage in lightweight design, prompting countries to continuously explore advanced preparation methods for rocket engine components using metal 3D printing technology.

Xi'an Bright Laser Technologies Co., Ltd. has implemented a typical part of an aerospace engine using metal 3D printing technology. The first layer is an adjustable nozzle structure (Fig. 3.46), which is one of the key basic parts that enable the engine to achieve good performance under different working conditions, increase engine thrust, and reduce fuel consumption. Based on 3D printing, the integration of the adjustable nozzle has become possible. During printing, the positioning structure of the nozzle and multiple sets of adjustment blade structures are integrated, achieving an adjustment angle of $\pm 20°$ at the engine nozzle. The second layer is a centrifugal compressor, which has a relatively simple structure and is widely used in medium and small-sized turbo engines and auxiliary power units in the aviation industry. It also uses 3D printing to achieve integrated design, with the wall thickness of the large and small blades being only 1.5 mm and 0.7 mm respectively. The third layer is a flame tube, an important part of the combustion chamber, which withstands extreme

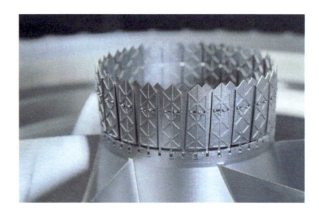

Fig. 3.46 Adjustable nozzle structure in 3D printed engine components. *Source* Xi'an Bright Laser Technologies Co., Ltd.

high temperatures and pressures during operation. Using 3D printing technology, the design of the cooling gas film structure of the flame tube has been realized, with the diameter of the gas film hole freely changing from 0.6 mm to 6 mm, enhancing design freedom. The fourth layer is a turbine guide vane, which is directly impacted by high-temperature and high-pressure airflow during operation, and has the function of changing the direction of the airflow, reducing the temperature of the airflow, and increasing the speed of the airflow. The fifth layer is the casing, which has the structural characteristics of a thin-walled shell and is the external protective device of the core machine, with a wall thickness of 1.5 mm. Using 3D printing, dot matrix skin reinforcement, topology reinforcement, and honeycomb reinforcement structures have been realized on the casing surface, not only reducing the weight of the parts, but also improving the rigidity and strength of the overall casing structure. At the same time, the integrated design reduces the number of parts and improves reliability, integrating the technical advantages and process characteristics of 3D printing with the performance requirements of engine components.

3.6 3D Printed Small Satellites

The booming technology of small satellites undoubtedly signifies the beginning of a new era in space exploration. They are known for their small size, light weight, low launch cost, and diverse functions.

CubeSats are a type of small satellite in a cube shape used to perform simple space observations and measurements of the Earth's atmosphere (Fig. 3.47). Due to their small size and weight, CubeSats have a much lower launch cost compared to larger satellites, but their functionality is limited and they are currently mainly used for school teaching, scientific research, and experiments.

Integrating 3D printing technology with small satellite manufacturing undoubtedly makes the production efficiency of small satellites extremely high and the cost extremely low, and provides a basis for printing small satellites in orbit. 3D printing

3.6 3D Printed Small Satellites

Fig. 3.47 CubeSats in Earth orbit. *Source* European Space Agency

technology makes the volume and shape of satellites variable, significantly reducing costs. The relatively simple structure of CubeSats makes them ideal research and experimental subjects for 3D printing technology.

Initially, CubeSats were manufactured using traditional spacecraft manufacturing technology. In recent years, due to the numerous parts of CubeSats, many external structures of parts have started to be made using 3D printing technology materials.

Our country has not only completed the first space experiment of 3D printing composite materials in orbit in the new generation of manned spacecraft test ship, but also carried the world's first 3D printed metal CubeSat deployer (Fig. 3.48). This CubeSat deployer, named COSPOD-3D, is used to accommodate multi-layer unfolding sail panels, large antennas, cameras, and other external protruding satellite payloads. COSPOD-3D was developed by Star Crowd Space (Beijing) Technology Co., Ltd. and manufactured by Xi'an Platinum Special Additive Technology Co., Ltd. based on metal 3D printing technology. Its weight is half that of traditional mechanically processed products, and the processing cycle has been shortened.

With the successful landing of the new generation manned spacecraft test ship return capsule in the scheduled area of the Dongfeng landing field in May 2020, the test ship flight test mission was successfully completed, which also verified the adaptability of the metal 3D printed CubeSat deployer in space.

Fig. 3.48 3D printed CubeSat deployer frame. *Source* Xi'an Platinum Special Additive Technology Co., Ltd.

Fig. 3.49 3D printed cubic satellite by Montana State University

Montana State University in the United States has prepared a cubic satellite using 3D printing technology, which is a case of 3D printing application in space (Fig. 3.49). The satellite uses a 3D printer and powdered polymer as raw materials to prepare the structure and system using 3D printing technology. The satellite system uses Windform XT 2.0 material, which is a composite material filled with polyamide and carbon, originally used for ground applications. The payload material did not use 3D printing technology, but used a single-chip composite radiation microdose meter as a force sensor, and the surface resistivity sensor measured the sensor performance of the satellite's nickel plating.

The Space Science and Engineering Laboratory at Montana State University designed the entire system using computer-aided design and manufacturing and other engineering tools, and tested the satellite. The satellite met NASA's technical standards, and Sandia National Laboratories prepared for flight. Unfortunately, due to a rocket launch failure, the 3D printed experimental satellite was unfortunately destroyed and did not obtain on-orbit operation data.

The European Space Agency has conducted tests using a new type of printable conductive hard plastic as raw material and completed the 3D printing of the cubic satellite structure. Once the components, circuits, and solar panels are installed, this 3D printed cubic satellite can be put into use at any time.

The 3D printing technology of polyether ether ketone (PEEK) material provides a basis for the printing of cubic satellite structures. PEEK, as a thermoplastic material, has excellent strength, stability, and temperature resistance, with a melting point as high as 350 ℃. By doping nano-particles into PEEK, this printable PEEK material will have conductivity, and then doped PEEK is used as the raw material for 3D printing filament (Fig. 3.50).

Based on this technological breakthrough, students from Delft University of Technology in the Netherlands carried out the printing work of the cubic satellite shell (Fig. 3.51).

The European Space Agency is conducting further research on this technology, studying the use of polyether ether ketone 3D printing technology in a microgravity environment to explore its potential future use in space. If realized, polyether ether ketone 3D printers will fly to space to serve astronauts in the space station, printing

3.6 3D Printed Small Satellites

Fig. 3.50 3D printed polyether ether ketone (PEEK) components

Fig. 3.51 3D printed cubic satellite shell

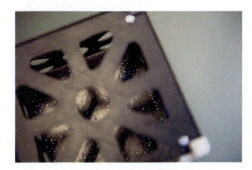

many useful tools and components. Especially by doping to give it conductivity, it could be one of the ways to manufacture small satellites and electronic components in space.

In August 2017, Russia released information about the world's first CubeSat, Tomsk-TPU-120, made by 3D printing, which entered the International Space Station aboard the MS-2 rocket (Fig. 3.52). Astronauts manually released 5 CubeSats during a spacewalk, the first of which had a shell and battery pack made by 3D printing. This CubeSat was manufactured by a student team from Tomsk Polytechnic University in Russia using 3D printing technology, measuring only 30 cm × 11 cm × 11 cm and weighing less than 4 kg. Despite its small size, the satellite is powerful, equipped with various sensors, capable of working in space for 4 to 6 months, and can transmit parameter information of various surface parts, circuits, and battery components of the satellite to Earth. Researchers can observe the working conditions and status of 3D printed materials in space based on this, and conduct feasibility analysis for its space adaptability and whether it supports further spacecraft manufacturing. In this process, the 3D printed CubeSat is exposed to Earth's gravity orbit decay and eventually burns up in the atmosphere after completing its work. Researchers believe that once these CubeSats are launched, they can form orbital clusters and even repair each other during operation, extending their lifespan.

Fig. 3.52 CubeSat Tomsk-TPU-120 made by 3D printing

Chapter 4
On-Orbit 3D Printing

For crew members who work long-term on spacecraft or space stations, they must carry everything needed for work and life in advance, which means allocating storage space for every possible item. For reserve supplies that are not needed in the short term, this undoubtedly wastes the very precious and limited storage space on the space station. At the same time, the carried supplies must be able to withstand the effects of gravity, vibration, and impact during launch, greatly increasing the cost of launch and ground testing. For longer space flights and deep space exploration missions, the unsustainability of the supply chain even becomes one of the important factors affecting the success of space missions due to time, distance, and duration.

Even on space stations in low Earth orbit, long-term space operations still require a lot of maintenance costs and logistical support. According to previous data, the International Space Station has reserved more than 13,000 kg of supplies to prevent failures. At the same time, more than 18,000 kg of supplies are stored on the ground for emergencies. These supplies have a low probability of use, and their storage consumes a lot of space resources and costs.

For resource replenishment in space stations, it currently mainly relies on satellite launches into orbit. Figure 4.1 shows typical cargo spacecraft and space freighters, used to supply the space station (Fig. 4.2) with necessities such as food, fuel, and experimental supplies.

On the basis of theoretical calculation, the cost of sending supplies to space is roughly equivalent to the value of gold of the same weight as the supplies. That is to say, to send 1 kg delivering supplies to low Earth orbit costs hundreds of thousands of RMB (tens of thousands of US dollars), and often the actual cost is many times higher. The cost of returning objects from space to Earth is even more expensive—the "Hayabusa" spacecraft brought back less than 1 g of asteroid particles, and the entire mission cost 250 million US dollars, equivalent to 2500 billion US dollars per kilogram of sample.

With the gradual development of 3D printing technology, especially the rise of 3D printing technology in space, all this will become a thing of the past. Necessary

Fig. 4.1 Cargo spacecraft and space freighters

Fig. 4.2 Space station

supplies and component reserves and replacements will all be carried out in space, which was unimaginable in the past. 3D printing in space will save a lot of costs and time, and promote the feasibility of space missions.

3D printing in space can not only achieve "on-demand production", greatly freeing up original storage space, improving space utilization and the actual use area of the spacecraft, but also "travel light" in deep space exploration and space travel, making long-term flights possible. In addition, the 100% utilization rate of materials and unlimited recycling and reuse are also reasons why scientists and aerospace engineers pay attention to this technology. On-orbit 3D printing shows people the infinite possibilities of technological innovation in space.

4.1 Challenges Faced by Spacecraft and Others in On-Orbit 3D Printing

On-orbit 3D printing will bring many benefits, greatly reducing the supplies needed to be carried in spacecraft, significantly reducing launch costs, and astronauts will no longer have the trouble of "item storage".

The "on-demand production" usage model will revolutionize the life and work modes in spacecraft. However, it should be noted that on-orbit 3D printing technology in spacecraft is different from on-Earth 3D printing technology, and still faces many technical challenges. There is still a long way to go before scientists and aerospace engineers can achieve their ideal "on-demand production".

The first and foremost is the complexity of the space environment. Vacuum, microgravity, cosmic radiation, extreme temperature changes, etc. are all issues that need to be considered, which will affect the printing process, material selection, material melting, model solidification and molding, etc.

For the vacuum environment in space, some 3D printing equipment is less affected, such as equipment that uses high-energy electron beams to melt materials, as the electron gun itself also works in a vacuum. But even so, the heat transfer, thermal conductivity, and material melting flow process in a vacuum are all different from the working state under normal pressure. On the one hand, we can partially eliminate the impact of the vacuum environment by choosing the appropriate 3D printing technology. On the other hand, we can improve the impact brought by the vacuum environment by adapting the printing equipment and raw materials. For some 3D printing technologies, such as photopolymerization molding technology, the vacuum environment has advantages such as no bubbles during the material molding process. As for the cabin environment, considering that the life and work of astronauts are often maintained within the normal pressure range, equivalent to the pressure at sea level on Earth. The cabin also simulates the atmospheric composition of the ground, mainly composed of oxygen and nitrogen, so there is no need to consider the impact of the vacuum environment. However, it should be noted that during the printing process, additional exhaust gases and toxic and harmful by-products should be avoided to prevent pollution of the cabin environment and pose a threat to the safety of astronauts.

Cosmic radiation mainly affects devices and equipment working outside the cabin. Whether it is material aging and brittleness caused by cosmic radiation, or space effects such as internal charging and discharging in devices, they will bring fatal harm to devices and equipment. For on-orbit printing work carried out outside the cabin and instruments and equipment that work outside the cabin for a long time, it is necessary to repeatedly test and verify the characteristics of the printing materials and the performance of the devices during the development process on Earth.

Extreme temperature changes are also a challenge for devices working and operating outside the cabin. For the interior area of a spacecraft, its temperature can be maintained at a comfortable level through various technologies and advanced equipment, facilitating the life and work of astronauts. However, for the outside of the

Fig. 4.3 Spacecraft in extreme temperature environments in space

(a) Spacecraft flying under solar radiation (high temperature)

(b) Spacecraft flying away from the sun (low temperature)

cabin, the spacecraft will face a wide range of temperature changes, possibly in the extreme temperature range of -157 to $+121$ ℃. This is mainly affected by solar activity (Fig. 4.3). Due to the lack of protection from the Earth's atmosphere, the spacecraft is exposed to space. When it rotates to face the sun directly, the surface temperature reaches about 121 ℃. When it rotates to the dark side of the Earth, where the sun is completely blocked by the Earth, the surface temperature drops to about -157 ℃. Therefore, when 3D printed materials or equipment are used on the exterior of the spacecraft or operate outside the cabin, the impact of extreme temperature environments must be considered.

The microgravity environment is one of the more important factors affecting 3D printing equipment and materials. Whether inside or outside the cabin, they are affected by the microgravity environment on orbit. When a spacecraft is in orbit, it is subject to various forces outside the Earth's gravity, including atmospheric drag, solar radiation pressure, etc. As a result, it does not reach a completely "weightless" state, but a "microgravity" state. Since the 3D printing process often requires the melting or jetting material to combine with the previous material under the action of gravity, the microgravity environment changes the process of material combination, and it is necessary to develop printing equipment specifically for the microgravity environment. In addition, the transfer of heat in the microgravity environment will also change, affecting the material forming, deformation, and scaling process.

Therefore, preliminary research and verification of microgravity 3D printing technology on the ground is essential. In this process, the following factors need to be considered:

4.2 Microgravity 3D Printing Equipment

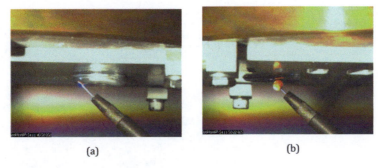

Fig. 4.4 3D printing in a microgravity environment. (a) The distance between the print head and the workbench is appropriate. (b) The distance between the print head and the workbench is unreasonable

① Control of the form of raw materials under microgravity conditions. Raw materials in the microgravity environment should be easy to manipulate and bond under melting conditions.
② The choice of 3D printing process under microgravity conditions. It needs to meet the restrictions of available space, equipment power supply, energy, high vacuum, cooling conditions, etc. in the space station/spacecraft.
③ Dynamic control during the melting and cooling process of raw materials under microgravity conditions. It should make the process of melting, forming, and solidifying of raw materials under microgravity conditions meet the predetermined requirements, and produce samples with the required precision. Figure 4.4 shows the situation of 3D printing in a microgravity environment using electron beam melting technology. When the printing head is at the right distance from the substrate on the workbench, it can print normally. When the distance between the print head and the substrate is too large, if it is in a gravity environment, the melted material will form droplets and fall on the substrate, while in a microgravity environment, the melted material forms a sphere at the end of the print head and then expands. If vibration occurs at this time, the droplets of the molten material will loosen in the vacuum chamber. They even float, leading to printing failure.

4.2 Microgravity 3D Printing Equipment

During the fourth phase of SpaceX's cargo resupply service mission, ground staff launched a space manufacturing 3D printer to the International Space Station, preparing for future parts to be printed on demand in space. NASA's Marshall Space Flight Center in Alabama and the Space Manufacturing Center in Mountain View, California, collaborated to develop and launch the first 3D printing device to the International Space Station.

Figure 4.5 is the 3D printer that SpaceX sent into space. It uses extrusion additive manufacturing technology, mainly through Acrylonitrile Butadiene Styrene (ABS)

Fig. 4.5 SpaceX's 3D printer sent to the International Space Station

Fig. 4.6 Plastic products printed in space

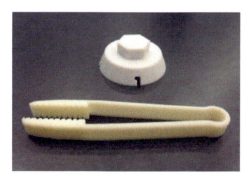

plastic (the same material used to manufacture Lego bricks) and other materials, layer by layer to print and build the objects to be printed (Fig. 4.6). With the computer-aided design files on the printer, this machine can print more than 20 parts and can transmit additional print files from Earth, printing in space via remote satellite communication.

Printing tools and equipment on demand in orbit greatly shortens the time and cost required for them to enter orbit, while improving the reliability and safety of space missions. Current space missions often require months to years to send parts into orbit. However, with on-orbit 3D printing, parts can be completed in a few minutes to several hours.

4.3 On-Orbit 3D Printing of Composite Materials

Space stations, also known as aerospace stations, operate for a long time in low Earth orbit, allowing multiple astronauts to patrol, work and live for a long time, and are also the first stop for people to carry out space experiments and explore space.

4.3 On-Orbit 3D Printing of Composite Materials

For astronauts on the space station, various life supplies are essential for long-term living on the space station. Currently, supplies are mainly transported by regular cargo spacecraft.

For the parts, tools, and materials needed in space station missions, it often takes weeks, months, or even years to wait for supplies through round-trip transportation. For future solar system exploration missions, such round-trip transportation supply methods are even more costly, the transportation process is more complicated, and it may even be impossible. Researchers from various countries have carried out preliminary research, striving to find new solutions.

On-orbit 3D printing technology is an important direction of technological development.

The on-orbit continuous fiber-reinforced composite material 3D printing carried out by China is the world's first known on-orbit continuous fiber-reinforced composite material 3D printing. As shown in Fig. 4.7, this composite material space 3D printer was independently developed by the researchers of the Aerospace Science and Technology Group Co., Ltd. and was launched into the predetermined orbit with China's new generation manned spacecraft test ship by the Long March 5B remote one carrier rocket. After the mission was completed, it successfully returned to the Dongfeng landing site with the return cabin of the new generation manned spacecraft test ship.

During the orbit, two continuous fiber-reinforced composite material samples were printed, including a honeycomb structure sample (for future lightweight design of spacecraft structures) and a logo of China Aerospace Science and Technology Corporation (CASC), as shown in Fig. 4.8.

Carbon fiber materials have long been widely used in spacecraft components, especially in the manufacture of antennas, with significant advantages of light weight and high strength. In the on-orbit fiber-reinforced composite material 3D printing, the carbon fiber filaments are continuous, which is beneficial to the performance improvement of composite materials. It has successfully verified the 3D printing technology of composite materials under microgravity conditions, providing a preliminary research basis for the future on-orbit preparation of parts, and even the on-orbit expansion of the space station.

Fig. 4.7 China's composite material space 3D printer

Fig. 4.8 3D printed continuous fiber-reinforced composite material samples

4.4 On-Orbit 3D Printing of Satellite Components

3D printing experiments in space show that aerospace agencies in various countries now have the preliminary ability to perform 3D printing under the microgravity environment of the space station. The microgravity 3D printer works normally in space.

Generally, the 3D printer nozzle extrudes heated plastic, or molten metal or other materials, and then manufactures layer by layer.

With the support of NASA's Marshall Space Flight Center, Space Manufacturing Company has developed a 3D printer that can operate under microgravity conditions, as shown in Fig. 4.9. The prototype of this printer is slightly smaller than the printer that the company will install on the International Space Station. Under the SBIR contract signed at the Marshall Space Flight Center, engineers from the Space Manufacturing Company began to build a machine model that is slightly affected by gravity. It also needs to be reinforced to survive the launch process. At the same time, any waste gas produced during the printing process that would contaminate a closed space environment like the International Space Station is not allowed. The machinery used in space must also consider the lack of natural convection due to the lack of gravity, which means that air must be circulated passively. In the early testing process, the effectiveness of the printer under microgravity was verified on an airplane flying in a parabolic arc, but it was limited by the short time and could not be fully verified.

This problem is being solved by launching a small microwave oven-sized microgravity 3D printer prototype to the International Space Station for space demonstration. There, it will create more than 30 objects, including calibration samples and tools such as wrenches. Most of the printing work will be based on model files loaded on the machine before launch, and other files will be uploaded to the printer from Earth during on-orbit work.

The Space Manufacturing Company has shown the final models of some products that its space 3D printer may produce on the International Space Station, as shown in Fig. 4.10.

Based on the printing capabilities of existing 3D printers, a list of tools, parts, and other items that are often lost or damaged on the space station can be determined,

4.5 On-Orbit Recycling of Materials

Fig. 4.9 Microgravity 3D printer of Space Manufacturing Company

Fig. 4.10 Space 3D printer and its printed models

including structural parts, torque tools, containers, clamps, and many other items. These items range from critical, complex to common. Existing research shows that about 30% of small parts and tools on the space station can be directly printed, while more complex large parts or tools may be assembled from printed small parts.

For astronauts, this is undoubtedly exciting news, as the waiting time for some simple parts will be reduced from months to minutes.

4.5 On-Orbit Recycling of Materials

For astronauts in the space station, the supply of materials, components, and parts is essential for daily work and life. However, considering the time and cost of transportation to and from Earth, if the recycling of materials in orbit can be achieved, it will undoubtedly bring great benefits to astronauts who work in orbit for a long time.

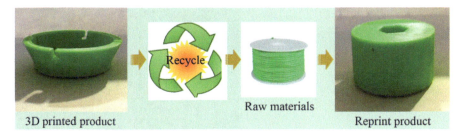

Fig. 4.11 Schematic diagram of on-orbit recycling of materials

Especially for long-term space flights, the recycling of materials in orbit is an important issue that scientists must consider. Compared with manufacturing on Earth and then launching into space by rockets, the recycling of materials in orbit presents a sustainable mode of space operation, providing a feasible logistics mode of manufacturing, recycling, and reusing for indispensable items that consume a lot of storage and transportation costs (Fig. 4.11).

Since NASA launched the first 3D printer into space in 2014, space manufacturing companies have collaborated with NASA again to launch the second 3D printer for on-orbit recycling of materials (Fig. 4.12). This printer, named "Refabricator", is mainly used to recycle waste parts and garbage on spacecraft, such as recycling the 10,000 sealed plastic bags that NASA has launched into space. The size of the printer is comparable to a small refrigerator, and it can print plastic materials of various shapes and sizes.

After this technology is researched and verified on Earth, it will continue to be verified in orbit in the space station, providing preliminary test and verification data for subsequent further long-term deep space voyages.

Fig. 4.12 Plastic recycling and reuse 3D printer. *Source* NASA

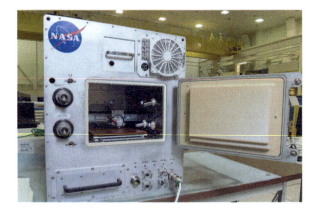

4.6 On-Orbit Construction and Assembly

The objects of on-orbit construction and assembly are generally those that need to serve in orbit in the space station, and the cost of manufacturing on the ground and then launching into space is high, or it is impossible to manufacture on the ground.

The proposal of on-orbit construction technology greatly expands the possible structural forms and sizes, freeing it from the limitations of launch weight and volume, and has infinite possibilities in the future.

For large-scale antennas and arrays, large-scale solar cell panels and arrays, space solar power stations, space telescopes, and other space science installation that are almost impossible due to the limitations of rocket carrying capacity or the impact of the launch process, the proposal of on-orbit construction technology has brought a glimmer of hope, providing a broader space for people to explore the universe.

In addition, in order to overcome the impact of strong vibrations and supergravity during the launch process, the structures and equipment launched into space must undergo special design and sufficient environmental tests on the ground, not only causing the development cost to remain high, but also greatly increasing the test cost and time cycle.

On-orbit construction overcomes these shortcomings, brings raw materials to space, and realizes the printing of functional structures with minimal raw material loss, eliminating the need for additional design to resist strong vibrations and supergravity during launching.

4.6.1 On-Orbit Construction 3D Printer Arm

When using 3D printing technology in space, astronauts often need to place the printing materials and operate the 3D printer. However, for on-orbit printing in extravehicular environments, such as large antennas and solar sails, scientists have begun to consider combining 3D printing technology with intelligent robot technology, proposing the concept of extravehicular on-orbit intelligent 3D printing.

Figure 4.13 shows the concept of an on-orbit 3D printing robotic arm. First, a 3D printer with a robotic arm is manufactured and installed on the exterior of a space station or spacecraft. The robotic arm is remotely controlled to manufacture and assemble large equipment and devices in space using 3D printing technology. At this time, astronauts do not need to operate directly, avoiding the dangers of extravehicular radiation and long working hours.

Currently, the combination of intelligent robots and 3D printing technology for large equipment on-orbit 3D printing and assembly technology is one of the most challenging problems in on-orbit 3D printing. The technical difficulties mainly lie in the integration of system engineering, robotic arm control system, printer control system, thermal analysis, software control platform, integrated testing, and other technologies. Once this transformative technology breaks through, the impact on

Fig. 4.13 Concept of on-orbit assembly and construction of solar cell arrays by a robotic arm 3D printer

human space exploration activities will be multifaceted and profound, bringing the following benefits:

① Ability to remotely construct communication antennas, large space telescopes, wireless energy collection and transmission systems, and other large complex systems in space.
② Enable small satellites to deploy solar sail arrays and antenna reflectors for medium and large satellites in orbit.
③ Eliminate the limitations on spacecraft volume and weight due to rocket carrying capacity.
④ Partially replace astronauts in performing extravehicular tasks, avoiding the inherent risks of spacewalks.

This technology is currently still being demonstrated and validated on Earth, simulating the temperature, pressure, and other conditions of the space environment, and printing related equipment or devices through a robotic arm. Subsequently, it is hoped to achieve on-orbit 3D printing in near-Earth orbit, providing feasible technical means for human exploration from the moon to Mars, and then to other planets.

4.6.2 On-Orbit Construction of Large Space Trusses

Trusses are structural components widely used in spacecraft, airplanes, radio telescopes, bridges, airports, and other large facilities and buildings (Figs. 4.14 and 4.15). The structural components must be able to withstand force loads from all directions and have vibration resistance, compression resistance, and other mechanical properties. For truss structures used on spacecraft, additional requirements include light weight, good mechanical performance, and good space adaptability, such as resistance to high and low temperature changes in extreme environments and radiation resistance.

Limited by the capacity of ground-to-air transportation and cost constraints, the conventional method of ground construction and rocket transportation for the preparation of large space equipment is clearly not feasible. For example, the size of a space solar power station can reach several kilometers after completion, and the weight can

4.6 On-Orbit Construction and Assembly

Fig. 4.14 Large truss structure in a radio astronomy telescope

Fig. 4.15 Large truss structure in an international airport

reach 2000 t or even more than 8000 t. Furthermore, for space super-large electric size antennas or space radio astronomy telescopes, to receive sufficiently weak cosmic signals, antennas with a diameter of tens of meters or even hundreds of meters are often required.

Large-scale solar cells and super-large antennas in space often use space trusses as support structures (Fig. 4.16). At this time, the on-orbit construction of space trusses becomes particularly important and is one of the important new technological development directions to solve the problem of constructing large-scale space infrastructure.

Currently, the construction method of large-scale satellite antennas of several tens of meters mainly involves the block manufacturing of truss structures on Earth, and then on-orbit assembly (Fig. 4.17); or the use of large-scale metal mesh structures, which are assembled on spacecraft after being constructed on the ground, and then expanded after the satellite is in orbit. The main technical barrier is the inability to cope with larger antennas, larger scale antenna arrays, and other applications. Currently, there are still huge technical difficulties and challenges in manufacturing antennas over 100 m and launching them into space for normal deployment, and it is even impossible to achieve.

Fig. 4.16 Large-scale solar cell array in space

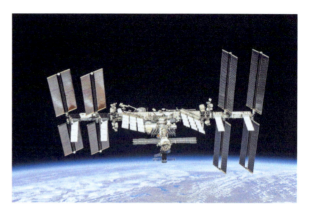

Fig. 4.17 Astronauts assembling in orbit

The on-orbit construction of space trusses is a completely new concept. By launching raw materials into space and 3D printing space trusses in space, only a small amount of assembly is required. The main technical advantage is that it not only breaks the limitation of the scale of space equipment construction by rocket carrying capacity, but also avoids the additional design needs caused by the vibration and hypergravity environment during the launch stage, and greatly reduces the cost and expense of rocket launch. In addition, it can also combine 3D printing for new structural design and weight reduction design, bringing new advantages combined with new technology (Fig. 4.18).

On-orbit 3D printing is considered to construct larger scale solar cell arrays, antenna arrays, trusses, etc., providing higher power, higher resolution, larger bandwidth, and higher sensitivity for in-orbit tasks of spacecraft.

Figure 4.19 shows a concept of on-orbit 3D printing construction of large-scale truss structures: extruding raw materials with good mechanical properties, layer by layer weaving, forming a "spider web" like space structure, which can increase the size of the space structure as needed. In this process, the following technologies need to be broken through:

4.6 On-Orbit Construction and Assembly

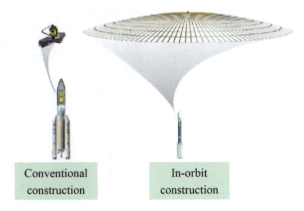

Fig. 4.18 Comparison of conventional construction mode and on-orbit construction mode

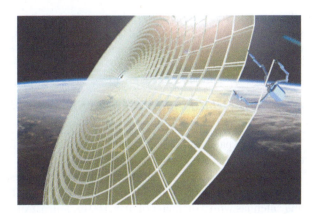

Fig. 4.19 Concept of on-orbit 3D printing of large-scale truss structures

① The molding method of the material forming the structure.
② The movement and control mechanism of the printer.
③ The combination and assembly method of the printed structure.
④ The thermal control method of the material and the formed structure.
⑤ The design and control of the printing process.
⑥ On-orbit integration technology of structural or functional components.

Interestingly, conducting 3D printing in space not only brings challenges but also unexpected benefits. When printing on the ground, due to the influence of gravity, on the one hand, the printed structure often inevitably has anisotropy, that is, the force is uneven in all directions, causing deformation or deterioration of mechanical properties; on the other hand, when structures overlap, are layered, or the print size is too large, additional support materials are often needed to resist the collapse or inability to form due to gravity. However, these problems will no longer exist during on-orbit construction.

On-orbit 3D printing can not only achieve uniform force in all directions, but also the print size can be as large as possible without being affected by overlap. This

Fig. 4.20 Concept of on-orbit 3D printing of ultra-large scale mesh antennas

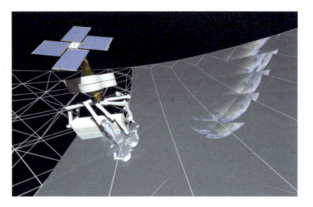

allows for the realization of ultra-large scale structural sizes and arbitrarily complex structural shapes without the need for additional support structures (Fig. 4.20), greatly expanding the design range and freedom of aerospace engineers.

4.6.3 On-Orbit Repair of Large Space Equipment

When large space equipment fails in orbit, the use of 3D printing technology for on-orbit repair shows broad application prospects.

Figure 4.21 is a concept for on-orbit repair of large space equipment. When large space equipment, such as ultra-large scale solar cell sails, space solar power stations, and antennas with a diameter of over 100 m, fail, astronauts initially explore the fault location and cause. Then, they scan the faulty part to complete preliminary analysis, and transmit the results back to the ground for data processing, analysis, and solution proposal. After that, the 3D printing model data of the part to be repaired is uploaded via remote satellite communication, and new parts or structural components are obtained through on-orbit 3D printing. Finally, the astronauts complete the on-orbit repair by assembling in orbit.

In this process, the exchange of data between the ground and on-orbit is particularly important. And on-orbit 3D printing technology is the key to the realization of on-orbit repair of large space equipment.

4.7 On-Orbit 3D Printed Satellites

A major application of on-orbit 3D printed technology will be the manufacture and deployment of nano-scale satellite shells, which can carry a number of new technology systems or experimental platforms for verification.

4.7 On-Orbit 3D Printed Satellites

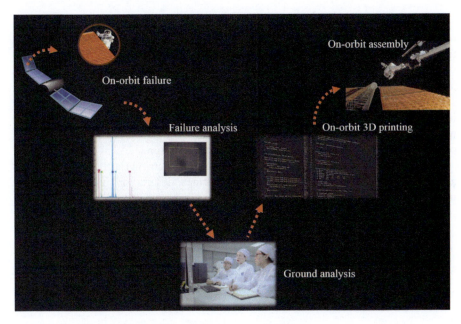

Fig. 4.21 Concept of on-orbit repair of large space equipment

Unlike the work mode of 3D printed small satellites on Earth and then launching them into specific Earth orbits by rockets, on-orbit 3D printing of small satellites can achieve the printing of satellite shells within a few hours. After inserting electronic circuits into them, the on-orbit printing and preparation of small satellites can be realized. The printed small satellite will be thrown directly outside the cabin, resulting in an on-orbit operating small satellite (Fig. 4.22).

Fig. 4.22 CubeSat

By launching many small packages containing raw materials into space via rockets, and then using a 3D printer installed on the space station, the preparation and deployment of micro-scale and nano-scale satellites in space can be realized, providing a more economical solution for various small satellite launch plans aimed at research and education. At this time, although the raw materials for making satellites still need to be launched and transported into space, the restrictions from logistics and size will be much less. More importantly, these satellites will not have to go through the harsh supergravity and strong vibration, during the launch stage, providing a more convenient solution for structural design.

This technology also has exciting application prospects on Earth. A durable, reliable 3D printer is particularly useful in submarines, deserts, or "anywhere you are in a remote area and rely on carrying parts with you".

Chapter 5
3D Printing in Deep Space Exploration

Compared to on-Earth and on-orbit 3D printing of spacecraft components, 3D printing technology in deep space exploration is particularly important and necessary due to its own technical advantages. As humans set off from Earth to step on the moon and gradually begin to explore further planets and deep space, 3D printing technology will play a key role in the supply of components during long-term deep space voyages and the construction of planetary bases. It is one of the most indispensable solutions among various technological approaches.

Planetary bases are necessary equipment for conducting planetary inspections, detecting resources, communicating with Earth, and providing astronauts with a place to work and live.

Imagine the following scenario: when we want to build bases on the moon (Fig. 5.1) or Mars, we need a large amount of building materials, which are often impossible to transport from Earth by spaceship. What should we do then? Making full use of the abundant soil and mineral resources on the planet and using 3D printing technology to build planetary bases in-situ is obviously the most economical and sustainable technical approach. Scientists from all countries are working hard for it. Some even regard the construction of lunar bases as the "second space race". Being the first to complete a stable, continuously operating lunar base is obviously very attractive to space agencies of all countries.

In addition, conducting research on 3D printing technology in deep space exploration is beneficial for people to assess the impact of the cosmic environment on humans, instruments, and spacecraft, and to provide the possibility of sustainable travel, exploration, and living in space. It will provide valuable technical information and basic data for space travel, the exploitation and utilization of extraterrestrial resources, and space refueling stations.

For long-term residence in space, the first thing to consider is the dependence on Earth and the independence of operation in space. The raw materials, tools, and equipment needed for living, production, and work during long residence in space should ideally be obtained independently as needed, without relying on Earth's supply.

Fig. 5.1 The moon as seen in space

Given the current technological development, 3D printing technology undoubtedly has broad potential applications in space. After all, for astronauts in space stations, especially for astronauts traveling in space, it is impossible and unnecessary to launch resource supply rockets from Earth whenever needed.

5.1 Challenges Faced by 3D Printing of Lunar and Other Planetary Bases

Space agencies around the world are actively exploring how to build bases on the moon and regard it as an important part of human exploration of deep space and the first step in exploring the solar system. 3D printing technology is undoubtedly a possible emerging technology for building lunar bases.

The process of building spacecraft landing sites, habitats, and runways on the moon is not as simple as on Earth. Accordingly, the construction process may also differ from the construction sites we commonly see on Earth. For example, excavation robots need to be lightweight and capable of excavating in a low-gravity environment. In addition, a large lunar base construction system often needs to be able to build autonomously, capable of carrying out construction work without the help of astronauts.

In this process, aerospace engineers will face various challenges. Using 3D printing technology for the autonomous construction of lunar bases is a very attractive and promising technical field. Of course, the challenges it brings are even greater.

Challenge One: Finding suitable materials for in-situ printing on the moon. When we build houses on Earth, we have used wooden structures and concrete blocks in the past, and now we often use reinforced concrete structures. When Eskimos build ice houses in the Arctic, they also need to stack ice blocks of the right size. When building a base on the moon, the first problem is to find suitable raw materials and process them into usable "bricks".

Challenge Two: Developing robots for building lunar bases. Once we have the raw materials and the bricks, we still need labor to stack the bricks. We can choose to stack layer by layer during the printing process, directly forming the building.

Fig. 5.2 Lunar surface environment

Or we can choose to first make bricks, and then use construction robots to stack the bricks. Whether it is stacking layer by layer directly in the printing process or using construction robots to build the base, it is necessary to transmit the printing instructions through remote satellite communication and solve the problems that may arise during the construction process. This is still a challenge for current technological development.

Challenge Three: Obtaining sufficient energy supply. The operation of 3D printers and construction robots cannot be separated from the supply of energy, which is the biggest problem facing current technological development. Scientists have proposed various ideas, and the more feasible methods currently include: ① solar power supply; ② reserve energy battery power supply; ③ wireless energy transmission from space solar power stations; ④ clean nuclear power supply; ⑤ lunar resource in-situ collection power supply. These power supply methods have their own advantages and disadvantages, and are the direction that scientists are actively exploring. Perhaps one day in the future, humans will discover a stable, clean, and sustainable energy source on the moon.

In addition, vacuum environment, lunar dust, lunar earthquakes, impacts of tiny meteorites, microgravity, especially extreme temperatures and other special environmental factors on the lunar surface (Fig. 5.2), as well as the unique environment of other planets, are all important factors testing the in-situ construction of planetary bases, and are also major challenges faced.

5.2 Where Does the Raw Material Needed for Printing Planetary Bases Come From

In order to reduce the number of launches and alleviate the burden of launch missions, using local materials for 3D printing of planetary bases can be said to be the most feasible sustainable planetary base construction technology, and it is also a technical problem that scientists from all countries are striving to solve. Especially for deep

Fig. 5.3 Lunar surface soil layer

space exploration that is more distant, transporting materials from Earth is obviously a huge project, costly and impractical.

For example, for the construction of a lunar base, by reasonably utilizing the lunar surface weathered layer, that is, the loose rocks and soil layer on the surface to form raw materials (Fig. 5.3), and then using 3D printing technology to build the raw materials into a planetary base, undoubtedly shows scientists a promising prospect.

However, before collecting the weathered layer on the moon for raw material preparation and 3D printing, sufficient preliminary research and technical verification must be carried out on the ground and in a microgravity environment (Fig. 5.4).

By transporting printing equipment and raw materials to the space station via space freighters, we can carry out the verification of 3D printing technology for weathered layer raw materials under microgravity. By conducting such experiments, we can verify the feasibility of constructing bases on planetary surfaces. The printed samples will be sent back to Earth for performance analysis and testing, and compared with ground samples, providing experimental data for the optimization and improvement of the technology.

Figure 5.5 shows the printer used for weathered layer 3D printing in the space station, consisting of a raw material extrusion nozzle and a printer bed, suitable for raw materials based on weathered layers. Conducting weathered layer 3D printing

Fig. 5.4 Docking of the Cygnus spacecraft carrying a microgravity environment weathered layer raw material 3D printing technology verification platform with the space station. *Source* NASA

5.2 Where Does the Raw Material Needed for Printing Planetary Bases Come From

Fig. 5.5 Printer used for weathered layer 3D printing in the space station

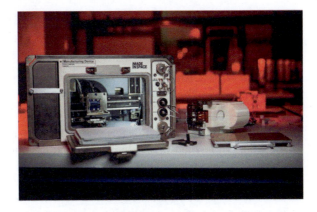

in space can verify and demonstrate the capability of 3D printing manufacturing process under microgravity, and produce material samples for scientific analysis.

On-orbit operations include installing hardware devices such as nozzles and machine beds for 3D printing, completing the printing of weathered layer samples, installing/removing RRP hardware (extruder and printer bed) from the device printer, remotely operating the printing of weathered layer samples and transmitting related data back to the ground via the downlink. The printed samples (Fig. 5.6) will eventually return to Earth, and all hardware used for printing will also be transported back to Earth for analysis.

Fig. 5.6 Weathered layer 3D printer components and printed samples

5.3 Latest Developments in 3D Printing Technology for Lunar Bases in Various Countries

5.3.1 China's Lunar Base Development Concept and 3D Printing Technology

In January 2019, China's National Space Administration announced that after the Chang'e 5 mission, the Chinese lunar exploration project will complete three more tasks (Chang'e 6, Chang'e 7, Chang'e 8), which will carry out sample return and comprehensive detection experiments in the lunar south pole. The lunar exploration project will also demonstrate the establishment of a lunar scientific research base.

The main purposes of establishing a base on the moon are: to better carry out scientific activities such as astronomical observations; to establish a space power station on the moon for use on Earth; to develop various mineral resources on the moon; to provide a foothold for human exploration further into deep space, etc.

The Chang'e 5 mission completed the collection of lunar samples, collecting about 2 kg of lunar soil in an unmanned manner. Figure 5.7 shows the lunar soil sample preserved in the Beijing Planetarium. Research on the lunar soil samples brought back by Chang'e 5 found that some components in the lunar soil can act as catalysts, converting water and carbon dioxide into oxygen and methane fuel under the action of sunlight. This may mean that in the future, lunar resources can be used to build lunar bases to support deep space exploration, research, and travel.

Over the past few decades, material scientists have analyzed lunar rocks and lunar soil samples and found that the moon contains abundant minerals and rare earth resources, including iron, aluminum, glass, rare earth elements, and silicon materials for manufacturing large solar panels. Most of the materials needed for lunar base construction can be found on the moon, which provides a feasible basis for "in-situ resource utilization" in lunar base construction.

Fig. 5.7 Lunar soil sample brought back by Chang'e 5

5.3 Latest Developments in 3D Printing Technology for Lunar Bases in Various Countries

Fig. 5.8 Simulated lunar soil prepared by the Institute of Geochemistry of the Chinese Academy of Sciences

The Chinese Academy of Sciences, China University of Geosciences, Jilin University, Tongji University and other research institutes and universities have all conducted research on lunar soil simulation. Figure 5.8 shows the CAS-1 simulated lunar soil developed by the Institute of Geochemistry of the Chinese Academy of Sciences using basalt lava collected from the Red Flag Forest Farm in Huinan County, Jilin Province to the Four Seas Forest Farm in Jingyu County as raw materials.

Building a base on the lunar surface is not an easy task. The imitation of lunar soil lays the foundation for obtaining 3D printing raw materials for lunar base construction on the ground. The key lies in how to use the resources available on the lunar surface for construction. Figure 5.9 shows the concept of a "cave dwelling" style lunar base. On the one hand, digging caves in lunar soil can use the thickness of the lunar soil to shield various radiations in the universe, save the cost of transporting building materials from the Earth, and solve the problem of rapid aging of building materials. On the other hand, this form of lunar base can make the most of the resources on the lunar surface for construction, reducing the difficulty of construction.

Fig. 5.9 Concept of "cave dwelling" style lunar base

5.3.2 NASA's Lunar Base 3D Printing Technology

As part of the Artemis program, NASA has proposed a concept for establishing the core surface elements needed for sustained existence on the moon. This concept emphasizes mobility to allow astronauts to conduct richer exploration and carry out more diverse scientific research. NASA is considering building lunar rovers, habitable mobile platforms, lunar RVs, and lunar surface habitats on the moon. The agency is investing in advanced manufacturing (one of the five industries for future space exploration and improving life on Earth), developing technologies that can find and use existing resources on the moon and Mars to build future infrastructure.

NASA is now working with ICON, a construction technology company based in Austin, Texas, to research and develop an early space-based construction system to support future exploration of the moon and Mars. ICON has built 3D printed communities of houses and buildings on Earth and participated in NASA's 3D Printed Habitat Challenge, demonstrating construction methods and technologies that may be suitable for off-Earth applications. Relying on NASA's MMPACT project, it has conducted research and testing on lunar soil simulation and various printing technologies. Governments, companies, and multiple research institutions are involved in the hope of improving the overall technology level and system testing capabilities to prove the feasibility of large 3D printers for building infrastructure on the moon or Mars. The team will use the knowledge and experience accumulated from lunar simulation tests to design, develop, and demonstrate a principle model of a full-size additive manufacturing system (Fig. 5.10).

In 2015, NASA launched a 3D Printed Planetary Base Challenge, hoping to use existing resources on the moon, Mars, or even further planets to build sustainable bases. The challenge includes modeling software, material preparation, and construction realization. The development of these new technologies will help people explore space and provide new solutions for low-cost housing construction on Earth.

Fig. 5.10 Conceptual diagram of lunar base infrastructure including lunar landing pads, bases, and other 3D printed facilities provided by ICON

5.3 Latest Developments in 3D Printing Technology for Lunar Bases in Various Countries

The 3D Printed Planetary Base Challenge is divided into three stages.

First stage: Design competition, requiring teams to submit architectural renderings, completed in 2015.

Second stage: Structural part competition, focusing on the development of material technology, requiring teams to realize structural parts, completed in 2017.

Third stage: On-site base competition, divided into five levels, testing the technical ability of each team to build a base, completed in April 2019. The base printing test is carried out monthly. Participating teams need to undergo tests in modeling software, material development, and construction, including two software design levels and three construction levels of competition. Among them, the sealing test is the final construction level that teams must pass. Only by passing the sealing test can they participate in the final level of competition.

In July 2018, in the first level challenge of the third stage of NASA's 3D Printing Planetary Base Challenge, 18 teams from around the world submitted their entries, as shown in Figs. 5.11, 5.12, 5.13, 5.14, and 5.15. After the selection by the expert panel from NASA, academia, and industry, five teams won a total of $100,000 in prize money for this level. They each used unique methods and specialized software tools to successfully create mathematical models of the physical and functional features of the planetary base.

In August 2018, in the second level of the third stage of NASA's 3D Printing Planetary Base Challenge, after evaluating all the entries, the following teams were rewarded: SEArch+/Apis Cor team, Pennsylvania State University team, FormForge/Austin Industries/WPM of Austin team. In this level of competition, each team needs to autonomously build base plates without human intervention. The expert panel tests and evaluates the base plates, then scores them based on strength, durability, and material composition. To test their strength, a standard Olympic shot put is thrown at each plate three times to simulate meteor impacts. The person in charge of the relevant aspects believes that this level of competition prepares each team for higher

Fig. 5.11 Zopherus team's 3D printed planetary base model

Fig. 5.12 AI.SpaceFactory team's 3D printed planetary base model

Fig. 5.13 Kahn-Yates team's 3D printed planetary base model

Fig. 5.14 SEArch+/Apis Cor team's 3D printed planetary base model

Fig. 5.15 Northwestern University Evanston Campus team's 3D printed planetary base model

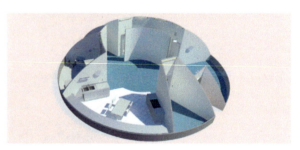

difficulty challenges, all construction must be completed autonomously, which adds extra difficulty to space exploration.

The SEArch+/Apis Cor team printed some base plates and withstood various tests, including using a blowtorch to simulate meteor impacts.

Figure 5.16 is the entry of the Pennsylvania State University team. In this photo, they throw a shot put onto a 3D printed foundation to simulate a meteor impact and test its strength. Figure 5.17 is the 3D printed entry of the FormForge/Austin Industries/WPM of Austin team.

In January 2019, in the third level sealing test of the third stage of NASA's 3D Printing Planetary Base Challenge, six teams submitted their entries. After the selection by NASA and the event organizer Bradley University, four teams won, namely the SEArch+/Apis Cor team, AI. SpaceFactory team, Pennsylvania State University team, and the Colorado School of Mines and ICON company team. In this level of competition, participating teams need to use existing resources on the planet to create a sustainable base suitable for the moon, Mars, or other planets. The participating teams submitted 3D printed samples and conducted sealing, strength, and durability tests under extreme temperatures.

Figure 5.18 is the 3D printed structure of the SEArch+/Apis Cor team, who won first place in the competition, preparing for a hydrostatic leak test. In the test, water

Fig. 5.16 Pennsylvania State University team's entry during testing

Fig. 5.17 FormForge/Austin Industries/WPM of Austin team's 3D printed entry

is injected into the 3D printed structure to a specified depth, and the leak rate is measured by the rate of water level drop. The Pennsylvania State University team won third place in the competition, and Fig. 5.19 shows the testing situation of their 3D printed structure.

In March 2019, during the fourth level of software modeling in the third stage of NASA's 3D Printed Planetary Base Challenge, NASA scored and ranked the entries of 11 teams based on the layout, planning, effective use of interior space, and the scalability and feasibility of 3D printing in the human living environment. Each team also prepared a short video explaining their design concept and a miniature 3D printed model for showcasing interior design. The aesthetics and practicality of the entries were also reflected in the scoring.

The SEArch+/Apis Cor team from New York won first place. This stage of the challenge required the participating teams to use modeling software to implement a complete and comprehensive base design plan. The base proposed by the SEArch+/Apis Cor team has a unique shape, can be continuously reinforced, and light can be projected through the slots on the side and top (Fig. 5.20).

Fig. 5.18 3D printed structure of the SEArch+/Apis Cor team preparing for a hydrostatic leak test

5.3 Latest Developments in 3D Printing Technology for Lunar Bases in Various Countries

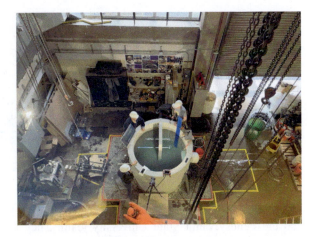

Fig. 5.19 Pennsylvania State University team conducting a hydrostatic leak test on their 3D printed structure

Fig. 5.20 3D printed lunar base model proposed by the SEArch+/Apis Cor team

The Zopherus team won second place. Their design consists of an automatic roaming 3D printer that prints out a structure and then moves to the next location (Fig. 5.21).

The Mars Incubator team won third place. The team, composed of engineers and artists, proposed a virtual design for a lunar base (Fig. 5.22).

Fig. 5.21 3D printed lunar base model proposed by the Zopherus team

Fig. 5.22 3D printed lunar base model proposed by the Mars Incubator team

The challenge was ultimately won by the AI. SpaceFactory team from New York and the Pennsylvania State University team. They built small-scale bases using resources and materials from the moon, Mars, and other celestial bodies, with the scale of these bases being 1/3 of the original design size. Each team used robots for construction, minimizing human intervention.

5.3.3 European Space Agency's Lunar Base 3D Printing Technology

The European Space Agency and the German Aerospace Center plan to establish a lunar simulation project at the European Space Agency's Astronaut Centre. Prior to this, the European Space Agency, in collaboration with industrial partners such as Foster + Partners, has already attempted to use 3D printing technology on Earth, using replicated lunar soil as raw material to print and build the bricks needed for a lunar base.

Researchers at the European Space Agency believe that obtaining raw materials from the moon and using 3D printing technology makes the construction of a lunar base possible and is the most feasible solution.

The research team tested the feasibility of preparing lunar soil and 3D printing. They first proposed a wall with a honeycomb structure for the construction of the base. This honeycomb structure wall adopts the bionic structure of bird bones, which has the advantages of high strength and light weight, and can be used to shield tiny meteorites and space radiation, and protect astronauts with a pressurized inflation device (Fig. 5.23).

The 3D printer of the honeycomb structure wall is provided by Monolite Company in the UK, with a D-shaped appearance and a frame size of 6 m (Fig. 5.24). The movable print nozzle array completes the printing by spraying adhesive solvent onto the sandy building material and stacking layer by layer. During the printing process, simulated lunar soil and magnesium oxide materials are mixed as the printing raw

Fig. 5.23 European Space Agency 3D printed lunar soil honeycomb structure wall. *Source* European Space Agency

Fig. 5.24 D-shaped 3D printer. *Source* European Space Agency

materials. The founder of the company said that the printer completes printing at a speed of 2 m/h.

To ensure that this 3D printing technology can be applied to the construction of lunar bases, it is necessary to solve the problems of evaporation and adhesion of the liquid material used in printing under vacuum. The research team members from the Italian Space Research Company and related universities used a method of inserting a 3D printing nozzle under the lunar weathered layer, so that the 2 mm diameter printing liquid can be bound by the capillary force of the lunar soil, ensuring that the printing process can be implemented in the vacuum environment on the moon.

Another problem the team needs to solve is the acquisition of lunar soil. Generally speaking, the composition of lunar soil is imitated and manufactured on Earth by professional companies, often used for scientific experiments. However, the team needs as much as several tons of lunar soil. In the end, they used basalt from a volcano in central Italy as a substitute for lunar soil, which has 99.8% similarity with lunar soil. In addition, further research on the impact of lunar dust and extreme temperatures on the lunar surface is needed for the 3D printing technology of lunar bases.

The European Space Agency also launched a lunar base 3D printing concept competition in 2019 (Fig. 5.25). The title of the competition is: To make the moon feel like home, what kind of 3D printing would you make on the moon? The competition received more than a hundred entries from adults and children around the world, including a mobile lampshade that can produce earth-like colors, an hourglass full of lunar dust, a real luminous earth glass model, suggestions for statues and game boards, and some suggestions about 3D printers. Two winners were finally selected, and their works are closely related to nature. Unlike NASA's lunar base competition focusing on technology and implementation, the European Space Agency's lunar base competition pays more attention to humanistic expression and concept establishment.

The adult group competition was won by British visual artist Helen Shell. She proposed the concept of "Magic Moon Garden", which is made of recyclable colored plastic 3D printing (Fig. 5.26).

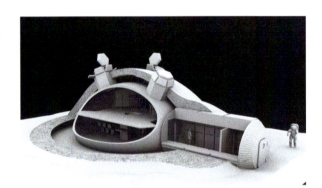

Fig. 5.25 European Space Agency 3D printed lunar base concept. *Source* European Space Agency

Fig. 5.26 Colored plastic 3D printed "Magic Moon Garden". *Source* European Space Agency

Judith De Santiago, a 17-year-old student from Madrid, Spain, won the championship in the under-18 youth group. She presented a dodecahedron plant pot that is very suitable for real plant cultivation, which symbolizes the distant earth.

5.4 3D Printing in Mars Base

5.4.1 *Mars Life Health and Exploration Mission*

In 2021, NASA solicited personnel for a one-year Mars simulation mission from the entire nation. This Mars simulation mission is named CHAPEA (Crew Health and Performance Exploration Analog). During the mission, various experiments will be carried out, and experimental data will be recorded, researched, and analyzed. The purpose of the mission is to provide possible solutions to potential problems for future manned lunar and Mars space missions.

The CHAPEA mission includes three one-year Mars surface simulations based at NASA's Johnson Space Center.

First phase: starts in the fall of 2022, recruitment began on August 6, 2021;

Second phase: start in 2024;

Third phase: planned to start in 2025.

The CHAPEA mission is carried out in the imitated Mars base at the Johnson Space Center. The base is named Mars Dune Alpha, and the main part is all completed by 3D printing. The four researchers participating in the mission will carry out all life, learning, and work activities in the Mars Dune Alpha, which is about 158 m^2 in area (Fig. 5.27).

In the 3D printed Mars Dune Alpha, challenges that may be faced in Mars missions will be considered, such as energy shortages, equipment failures, communication delays, and other adverse effects caused by the environment. Researchers in the base will carry out a series of scientific experiments and research work, such as space-walking, scientific research, the use of virtual reality, robot control, and communication exchanges. The results of the research will provide potential solutions and

Fig. 5.27 Concept of 3D printed Mars Dune Alpha base

important preliminary scientific data for the system verification of the Mars base and the problems that may be encountered in future Mars exploration, providing valuable first-hand experience and information for researchers to carry out Mars exploration.

5.4.2 Mars Dune Alpha

The 3D printed structure known as Mars Dune Alpha will be used as a substitute for a real Mars habitat to support long-term exploration-level space missions and provide similar expected experience for future Mars surface habitats. The layout of the 3D printed habitat is designed to provide separate areas for living and working within the habitat (Figs. 5.28 and 5.29).

The American company ICON successfully printed a single-story house up to 2.6 m high, 8.5 m wide, and almost unlimited in length on the ground based on its Vulcan series of architectural printers. It takes about 24 h to build a single-story house of about 32 m^2.

With the support of NASA, ICON collaborated with architectural groups such as Bjarke Ingels Group (BIG) to explore the construction of a 3D printed Mars base.

Fig. 5.28 3D printed Mars Dune Alpha

Fig. 5.29 3D printing process of Mars Dune Alpha

5.4 3D Printing in Mars Base

Mars Dune Alpha is printed from lava concrete and covers an area of about 158 m^2 (equivalent to the size of a three-bedroom and one-living room residential house).

In the mission of CHAPEA, several members are selected to live in the 3D printed Mars Dune Alpha for a 1-year phase-one mission. Each group will consist of four official members and two alternate members. In the Mars Dune Alpha, the mission will evaluate NASA's space food system and the physical, behavioral health, and various outcomes of astronauts when carrying out future space missions, providing extremely valuable data and information for the actual execution of Mars exploration missions. Research from the Mars Dune Alpha habitat will be used by NASA to predict risks and anticipate resource information, and to assess the health and performance of astronauts living on Mars during long-term missions.

The 3D printed Mars Dune Alpha mainly includes:

① Four astronaut dormitories;
② Dedicated workstations;
③ Dedicated medical stations;
④ Public rest area;
⑤ Kitchen and food cultivation station.

5.4.3 In-Situ 3D Printing of Mars Base

Although countries around the world are developing space technology in hopes of successfully landing on Mars, there is little discussion about living on Mars. Recently, experts have pointed out that "building houses" on Mars should use local materials, using 3D printing technology to synthesize the flying sand on the surface of Mars into appropriate building materials, so there is no need to transport building materials by spacecraft (Fig. 5.30).

Lunar bases or Mars trips require huge spacecraft, where fuel occupies a large space. If building materials for establishing a base need to be transported from Earth, space will be even more limited. Therefore, the space bases envisioned by scientists in the past were often made of extremely light and thin materials, as long as they

Fig. 5.30 Concept of planetary base

prevent leaks and resist cosmic radiation. However, using such light and thin space building materials on Mars poses a great risk. Due to the sandstorms on Mars, the existing materials are too thin to resist, and robust materials are still in need.

For this reason, similar to the 3D printing technology of the lunar base, scientists have proposed a new concept of in-situ 3D printing of Mars base, directly using the most abundant materials on the surface of Mars as building raw materials. The surface composition of celestial bodies like the Moon and Mars is actually similar to that of Earth, mainly a mixture of alumina, silica, iron oxide, and other minerals.

Therefore, researchers have conducted experiments using artificial lunar and Martian soil samples. The research shows that the soil on the lunar surface is often sharp and jagged due to the lack of air, and the grains are very rough. While the flying sand on Mars is more eroded by wind, and its shape is closer to the round grains on Earth. Therefore, researchers consider the universality of the two types of sand when making 3D printers, in order to design a more versatile space base 3D printer. By combining lunar and Martian soil with organic solvents, a printable or stackable paste-like building material can be prepared, and then different shapes of bricks can be stacked using a 3D printing nozzle. Finally, these bricks can be assembled together to form a sturdy long-term base, similar to the process of assembling building blocks.

5.5 3D Printing in Deep Space Exploration Instruments

In deep space exploration activities, in addition to establishing planetary bases, people are also committed to the exploration of planetary environments and the development of planetary resources. In the process of deep space exploration, instruments and equipment such as rovers and exploration robots are involved, and the application of 3D printing technology has gradually emerged.

Figures 5.31 and 5.32 respectively show the working scenes of China's Zhurong Mars Rover and NASA's Perseverance Mars Rover. Through the Mars Rover, it is possible to explore whether there are signs of microbial life on Mars, and also to collect information related to Mars' topography and geological structure, surface soil characteristics and water ice distribution, surface material composition, atmospheric ionosphere and surface climate and environmental characteristics, Mars' physical field and internal structure, etc., laying the foundation for further Mars exploration.

The use of 3D printing technology could realize the processing and preparation of various planetary exploration components and instruments. As shown in Fig. 5.33, the heat exchanger is used for the in-situ utilization of oxygen resources on Mars. The in-situ production and utilization of industrial oxygen on Mars can provide raw materials for the manufacture of rocket propellants on Mars, helping astronauts return to Earth.

One of the most difficult things when sending astronauts into space for planetary exploration or interstellar travel is to return to Earth. To return to Earth from the surface of Mars, a large amount of oxygen will be needed to synthesize propellants

5.5 3D Printing in Deep Space Exploration Instruments

(a)

(b)

Fig. 5.31 Mars rover working scene. (a) Zhurong Mars Rover. *Source* China National Space Administration. (b) Perseverance Mars Rover. *Source* NASA

Fig. 5.32 Mars Rover

Fig. 5.33 3D printed nickel alloy heat exchanger

with rocket fuel. For example, a crew of four needs about 25 tons of oxygen to generate about 7 tons of rocket fuel thrust. Therefore, the in-situ production of oxygen on the planet is crucial for the safe return of astronauts from deep space exploration.

In traditional manufacturing processes, the various components of the heat exchanger need to be processed separately and then connected together by welding. However, nickel alloy 3D printing could be used for the implementation of integrated design and manufacturing. The reason for using nickel alloy as the raw material is that nickel alloy is a high-temperature alloy that can maintain strength at very high working temperatures. Inside the heat exchanger, it must be ensured that the Martian air can be heated to an environmental temperature of about 800 ℃, and the nickel alloy will ensure that the key parts of the instrument are not affected by high temperatures.

This high-temperature alloy can also be used in jet engines or power turbines, with corrosion resistance and high-temperature resistance characteristics.

However, during the 3D printing process, as the device is formed layer by layer, pores or cracks may form between the layers, reducing the structural strength of the device. To deal with this situation, the formed material needs to be post-processed in a high-temperature and high-pressure environment to ensure its mechanical properties. The X-ray is used to detect the internal defects and cracks of the 3D printed nickel alloy heat exchanger in Figs. 5.33 and 5.34, and the final product presents a good processing state.

Figure 5.35 shows the shell of a Mars rover made by metal 3D printing. Compared with traditional mechanical processing technology, the 3D printed titanium alloy shell reduces its weight to only 25%–20% of its original weight. It presents a hollow structure with a thin shell. For the ray chemical rock detection instrument in the Perseverance Mars Rover's X, multiple 3D printed parts are used, which have the advantages of reducing weight, reducing the number of parts, and improving mechanical performance.

By launching 3D printed instruments or devices containing 3D printed parts into deep space environments or planetary surfaces, and observing the changes in their performances, indispensable data and information will be provided for further process optimization and space adaptability research.

Fig. 5.34 X-ray image of 3D printed nickel alloy heat exchanger

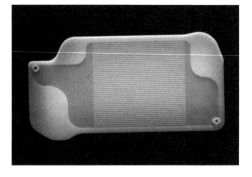

Fig. 5.35 3D printed Mars rover shell

More than 100 3D printed parts were also used in China's Tianwen-1 Mars rover. In addition to the improvement of radiation resistance, strength, corrosion resistance, and high-temperature resistance, structural topology optimization design methods were used for the reduction of weight, providing preliminary test references for the normal operation and running of 3D printed detection instruments and equipment in future planetary extreme environments.

5.6 New Materials Suitable for Deep Space Exploration Under Extreme Temperature Environments

For deep space exploration missions, not only astronauts, but various space equipment also undergo the harsh test of the space environment. Whether it is the establishment of a planetary base or a planetary exploration robot, there is a demand for materials and their preparation methods that can withstand extreme temperature environments.

For example, for robots used for planetary exploration, gears are the most important moving joints. Precise gears are crucial to robots and are the prerequisite for maintaining stable limb movement. However, for deep space exploration, extreme temperature environments may cause the lubricant of the gear to fail, the gear cannot work normally or even freeze and crack, or the service life is extremely shortened under high and low temperature changes. For the Curiosity Mars rover, it takes nearly 3 h to heat the lubricant before each start of work, consuming a lot of precious battery energy.

Metallic glass belongs to a class of materials with a unique disordered atomic structure, which combines the good ductility of metal, hardness exceeding highhard tool steel, and the toughness of glass, while also being corrosion-resistant and radiation-resistant. It could work at low temperatures of -200 ℃, and is an excellent emerging aerospace material. During processing, the metallic glass is blow molded by heating, so it is suitable as a raw material for 3D printing. When combined with

Fig. 5.36 Metallic glass gear working at extremely low temperatures

customized printing equipment, it leverages the advantages of metallic glass to make devices.

The gears made of metallic glass can achieve strong torque and smooth turning at −200 ℃ without lubricant (Fig. 5.36). For Mars exploration robots, using this kind of gear reduces the dependence on lubricants during movement, no longer needs extra power to heat the lubricant, and reduces battery power consumption.

Since 2017, there have been published patents for 3D printing technology based on metallic glass, with inventors from the California Institute of Technology. The principle of making metallic glass at the California Institute of Technology is similar to direct energy deposition technology: the first layer of metal alloy surface is melted at high temperature. Then this layer of molten metal alloy is quickly cooled and solidified to form the first layer of metallic glass. After that the next layer is processed on this basis.

The Jet Propulsion Laboratory has proposed a concept of using a metallic glass material for the fabrication of gears in extreme temperature environments. This material will greatly benefit the movement of robots in harsh environments, such as on Jupiter's moon Europa. The gears made of this special material are integrated into various parts of the robot arm, planned for lunar surface tests in the next few years (Fig. 5.37). After the arm assembly is completed, each joint containing the metallic glass gears needs to be tested, including torque, speed, low-temperature thermal vacuum tests, etc., to assess its real performance in space.

Fig. 5.37 Robot joint containing metallic glass gear motor after integration

With the gradual application of metallic glass 3D printing technology in the future, it may change the traditional manufacturing and movement methods of robots in deep space exploration missions, making it possible for robots to work efficiently on the moon, Mars, or even harsher space environments.

Chapter 6
New Materials for Space 3D Printing

6.1 3D Printing of Metallic Materials

6.1.1 Heterogeneous Dual Metal 3D Printing

The rocket engine igniter, used to initiate the engine's start sequence, is one of the complex components made from different materials. In traditional manufacturing industry, the igniter is made using a process called brazing, which connects two types of metals by melting, forming a bimetallic component. The brazing process requires a lot of manual work, leading to high costs and extended manufacturing procedure.

Through eliminating the brazing process and building the bimetallic components in one machine, it not only reduces costs and shortens manufacturing cycles, but also lowers risks and improves reliability (Fig. 6.1). By diffusing the two materials together, they bond internally and avoid the cracking that might occur due to the strong mechanical and temperature gradients during space travel (Fig. 6.2).

For the igniter prototype, two metals (copper alloy and chromium–nickel–iron alloy) are connected using a unique mixed 3D printing process called automated powder laser deposition. The igniter prototype is manufactured as a single component, rather than the four different components that were previously brazed and welded together. This bimetallic component is manufactured in a single process using a DMG/MORI hybrid machine, which integrates 3D printing and computer numerical control machining capabilities.

Although the igniter is a relatively small component, only 10 inches (about 0.25 m) high and a maximum diameter of 7 inches (about 0.18 m), this new technology allows for the manufacture of larger parts and enables the interior of the parts to be machined during the manufacturing process, something other processing technologies cannot achieve. It likes building a ship inside a bottle, where the exterior of the component is the "bottle", containing a detailed, complex "ship" with unseen details. The hybrid process can work inside the component before the exterior is completed and closed

Fig. 6.1 3D printed bimetallic lightweight thrust chamber components at NASA's Marshall Space Flight Center

Fig. 6.2 Microscopic images reveal the strong bonding structure produced by 3D printing through the mixing and interlocking of two metals (copper alloy and chromium–nickel–iron alloy)

alternating freely between 3D printing and machining. In the next generation of rocket engines, it is expected to create larger, more complex flight parts through 3D printing technology.

The Robotic Deposition Technology (RDT) team at NASA's Marshall Space Flight Center is designing and manufacturing new lightweight combustion chambers, nozzles, and injectors using automated robotic deposition 3D printing technologies: cold spray deposition, laser wire direct closeout, laser powder bed fusion, and laser powder directed energy deposition. This hardware has been started 8 times in total during a total hot fire duration of 365.4 s. In all ongoing tests, the pressure in the main combustion chamber is as high as 750 psi (5.17 MPa), and the calculated hot gas temperature is close to 6200 ℉ (3427 ℃). Three different carbon fiber composite nozzles designed for 31,137 N of thrust were also tested, demonstrating their ability to withstand extreme environmental conditions, with measured nozzle temperatures exceeding 4000 ℉ (2204 ℃). The progress of RDT will benefit future space and commercial space missions, as it provides lighter and cost-effective liquid rocket engine parts, rather than traditional hardware, which is heavier and usually composed of more parts.

6.1.2 3D Printing of Large Copper Parts

Copper has excellent thermal conductivity, which is why it is the ideal material for making engine combustion chambers (Fig. 6.3) and other parts, but this characteristic makes additive manufacturing of copper challenging, as lasers struggle to continuously melt copper powder.

Currently, only a handful of copper rocket parts are made using additive manufacturing, so NASA is pioneering a new technical field by 3D printing a rocket part that must be able to withstand extreme high and low temperatures, and establish complex cooling channels on the very thin outer wall. The part is made of GRCop-84, a copper alloy manufactured by NASA's Glenn Research Center. Extensive material characterization has helped validate the 3D printing processing parameters and ensure manufacturing quality. The Glenn Research Center will develop a comprehensive mechanical property database to guide future 3D printed rocket engine designs.

Inside the combustion chamber, the propellant burns at temperatures exceeding 5000 °F (2760 °C). To prevent the part from melting, ultra-low temperature hydrogen is circulated in over 200 finely crafted cooling channels.

The electron microscope image in Fig. 6.4 shows the original copper powder used to build the 3D printed copper liner. Scientists characterized samples of this alloy (Fig. 6.5) to understand how powder quality and characteristics affect component quality.

In addition, a selective laser melting printer in the Materials and Processing Laboratory at the Marshall Space Flight Center melted 8255 layers of copper powder to make the combustion chamber in 10 days and 18 h. Before manufacturing the liner,

Fig. 6.3 Engine combustion chamber printed entirely in copper

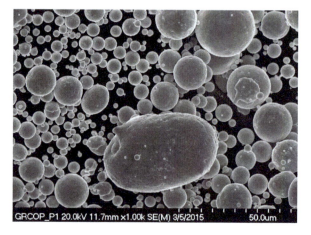

Fig. 6.4 Electron microscope image of original copper powder

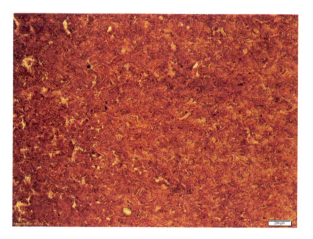

Fig. 6.5 Optical microscope photo of printed all-copper sample

material engineers made several other test parts, characterized the material, and created a process for additive manufacturing with copper.

6.1.3 *3D Printing of Iron Nickel Metal*

Through 3D printing (Fig. 6.6), the final cost of complex, performance-optimized, lightweight parts is lower than traditional alternatives. The AMAZE project (Additive Manufacturing Aiming Towards Zero Waste and Efficient Production of High-Tech Metal Products) has been developing 3D printed parts to withstand the required load, ultimately making these performance-optimized complex metal parts weigh less than half of the original components, and they are made in one piece, eliminating the typical potential weak points at the joints.

Fig. 6.6 Laser-based 3D printing

AMAZE was launched in 2013 and is the largest 3D printing R&D program in Europe to date. The program produces lighter, cheaper, and better-performing parts. AMAZE's work covers the entire process chain, including proposing new methods for part design, reliably completing and inspecting the parts produced, introducing new materials, increasing production volume, and setting common industrial standards.

For the meter-scale printed parts in Fig. 6.7, the complexity of these parts means that the size of the print model file could be huge (several orders of magnitude larger than a normal CAD file), and it may take a long time to process all the data. Therefore, another important direction for 3D printing is the development of new software tools to fundamentally reduce the required time.

The development of new materials is to meet specific industrial needs, including the first 3D printing of in var, a nickel-iron alloy, which is highly valued by the European Space Agency for its ability to withstand extreme temperatures in space orbits without expanding or contracting (Fig. 6.8).

Another issue for printed parts is the post-processing, finishing, and inspection of the parts, including standardized non-destructive testing procedures. For example,

Fig. 6.7 Meter-scale titanium cylinder printed in 3D space

Fig. 6.8 High-temperature resistant in var nickel-iron alloy printed parts

Fig. 6.9 3D printed tower bracket

the tower bracket completed by 3D printing in Fig. 6.9, the structural accuracy of its machined parts needs to be strictly matched, especially for the microstructures. Therefore, the use of medical-style three-dimensional CT scanning is a solution being explored for the precision inspection of machined parts, and existing research results are preparing for the establishment of a universal ISO standard in this field.

6.1.4 3D Printing of Titanium Metal

Laser melting technology is used to achieve 3D printing of titanium alloys, such as the test version of the optical workbench core of the European Space Agency's Athena X-ray observatory (Fig. 6.10). A multi-axis robotic arm is being used to produce complex structures, including structures for placing optical mirror modules. This is achieved through a specially developed process called COAXShield, which uses the rare gas argon to sweep titanium powder into the path of the laser, protecting the newly printed titanium from contact with the atmosphere during this process.

6.1 3D Printing of Metallic Materials

Fig. 6.10 European Space Agency using laser melting technology for 3D printing of titanium alloys

This gas protection allows for a quick switch between additive manufacturing (laser metal powder deposition) and subtractive manufacturing. Considering that the low-temperature cooling milling tool operated by the second robotic arm could remove excessive titanium, the optical workbench itself is placed on a slowly moving 3.4 m-diameter turntable between the two robotic arms. The ultimate goal of the project is to produce a 3 m-sacle optical workbench. It is predicted that the technology of this program can be applied to other programs for the printing of various sizes.

6.1.5 3D Printing of Metal Ceramic Parts

TIWARI Scientific Instruments, a German start-up supported by the European Space Agency, has developed a technology that allows low-cost 3D printing with various metals and ceramics. Typically, producing precision parts with such high-performance materials is expensive in terms of time and cost. However, the manufacture of precision parts could be executed using standard 3D printers through this technology. These parts have been made with alumina ceramics, which meet space quality standards using only a ready-made desktop 3D printer.

TIWARI's Fused Filament Fabrication (FFF) printing process uses thermoplastic filaments embedded with metal or ceramic particles (Fig. 6.11). After printing, the part, known as a green body, will undergo heat treatment to remove the plastic, leaving a metal or ceramic item.

6.1.6 Stainless Steel 3D Printing

The bottle opener in Fig. 6.12 is 3D printed with stainless steel, meeting space standards, using only a ready-made desktop 3D printer. This part is also manufactured using the aforementioned FFF printing process. Once the printing is completed, the part will undergo heat treatment to remove the plastic, leaving a metal or ceramic item.

Fig. 6.11 Alumina ceramic 3D printing based on fused filament fabrication

Fig. 6.12 3D printed bottle opener

6.2 3D Printing of Composite Materials

6.2.1 On-Orbit Recycling and Reuse of 3D Printed Plastics

The first integrated 3D printer and recycler was part of the cargo launched to the International Space Station during the 10th commercial resupply service mission of the Cygnus spacecraft by Northrop Grumman. This machine, known as the Refabricator, demonstrated the ability to turn waste plastic and previous 3D printed parts into high-quality 3D printer filament (3D printing "ink") to manufacture new tools and materials. Parts made from recycled plastic multiple times have been quality tested on Earth. It has been proven to aid future lunar and Mars exploration missions. The device completed its final flight certification test at NASA's Marshall Space Flight Center.

6.2 3D Printing of Composite Materials

Fig. 6.13 The Refabricator

The Refabricator accepts plastic materials of various sizes and shapes and turn them into raw materials for 3D printing items. The entire process is completed in an automated machine about the size of a dormitory refrigerator (Fig. 6.13).

The Refabricator is a recycler and 3D printer, located in a machine about the size of a dormitory refrigerator, launched to the space station in April 2018.

Considering that it is impractical to send replacement parts or tools for all materials on spacecraft, and the cost and time of resupply from Earth are high, the Refabricator will be key to a sustainable logistics model for manufacturing, recycling, and reusing parts and waste. Astronauts can use this technology to manufacture and recycle food utensils, and transform waste that is currently inconvenient to handle into raw materials to help establish the next generation of space systems. Reusing these wastes can reduce the cost and risk of space missions and private space exploration missions.

6.2.2 Printing Bricks with Artificial Lunar Dust

As shown in Fig. 6.14, it is a triangular brick printed from artificial lunar dust using concentrated sunlight. It was produced using a 3D printer, continuously baking a 0.1 mm layer of simulated lunar dust at 1000 ℃, completing a 20 cm × 10 cm × 3 cm construction brick in about 5 h. The raw material for this experiment was commercially available artificial lunar soil based on terrestrial volcanic materials, processed to be similar to the composition and particle size of real lunar dust.

From the cross-section in Fig. 6.15, some bricks show warping at the edges because that they cool faster than the center. This effect can be alleviated by speeding up the printing process, thus reducing the amount of heat accumulated within the brick.

In the solar furnace (Fig. 6.16) at the German Aerospace Center in Cologne, 147 curved mirrors focus sunlight into a high-temperature beam, melting soil particles together. Considering affecting factors caused by different weather, sunlight is simulated by an array of xenon lamps. The resulting bricks have a strength equivalent to plaster and have undergone detailed mechanical testing. Nevertheless, at present, this

Fig. 6.14 Brick printed from artificial lunar dust using concentrated sunlight

Fig. 6.15 Cross-section of the printed brick

project is still a proof of concept, demonstrating that such a method of constructing lunar soil bricks is feasible. The experiment was conducted under standard atmospheric conditions, but the real lunar environment also faces the challenges of vacuum and extreme temperature conditions.

6.2.3 *Printing Concrete with Urea and Artificial Lunar Dust*

As shown in Fig. 6.17, the mixture printed by a 3D printer using a high-pressure injection pump is more solid and maintains good workability. Researchers found that adding urea to the lunar geopolymer mixture (a building material similar to concrete) works better than other common plasticizers (such as naphthalene or polycarboxylate), reducing the need for water. The mixture printed by the 3D printer is more solid and maintains good workability.

Figure 6.18 is a 1.5 t hollow structure, 3D printed using simulated lunar dust, enabling it to combine strength with low weight like a bird's bone. Produced using

6.2 3D Printing of Composite Materials

Fig. 6.16 Solar furnace at the German Aerospace Center

Fig. 6.17 High-pressure injection pump used for 3D printing of lunar dust simulant

salt as the "ink", this structure was manufactured in the initial feasibility project of lunar 3D printing.

Fig. 6.18 Hollow structure printed with simulated lunar dust

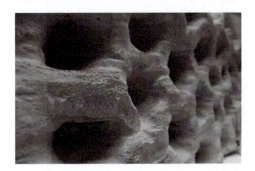

Fig. 6.19 Coin-sized 3D printed silicon carbide filter

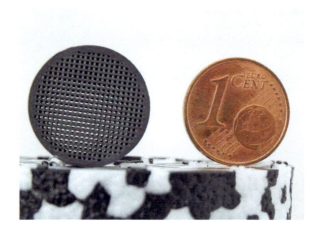

6.2.4 Printing Filters with Silicon Carbide

The size of the filter in Fig. 6.19 is comparable to a coin, made of silicon carbide, and meets space quality standards. Its production only used an off-the-shelf desktop 3D printer. This silicon carbide filter also uses the FFF technology developed by the German company TIWARI, implemented with standard 3D printing technology.

6.3 3D Printing of Biomimetic Materials

6.3.1 3D Printing of Bones and Skin

3D printing of human tissues helps astronauts stay healthy. A project led by the European Space Agency has produced the first batch of bioprinted skin and bone samples (Fig. 6.20). The bone sample uses human plasma as a nutrient-rich "bio-ink", with the addition of calcium phosphate bone cement as a structural support material, plus methyl cellulose and alginate from plants and algae to increase the viscosity of this "bio-ink", making it suitable for use in low gravity conditions.

The fabrication of artificial bone sample is a preliminary step in 3D bioprinting, which is a practical tool for space emergency medicine. A research and development plan raised by the European Space Agency aims to develop bioprinting technology that allows astronauts performing long-term missions to have access to the "essential parts" needed for bone or skin transplants, or even complete internal organs. 3D bioprinting is promising to meet the challenging conditions during space flight. For example, astronauts lose bone density in zero or low gravity, making fractures more likely to occur on orbit or Mars. Burn treatment often involves skin grafts taken from

Fig. 6.20 3D printed human bone sample

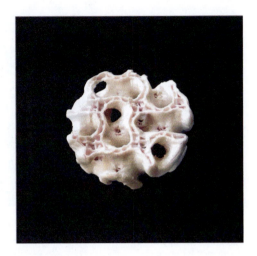

the patient's own body, which can be handled with full hospital care on Earth, but the risk is greater in space because secondary injuries are also not easy to heal.

6.3.2 3D Printing of Blood Vessels

Two teams of scientists from the Wake Forest Institute for Regenerative Medicine in Winston-Salem, North Carolina, took first and second place in the Vascular Tissue Challenge organized by NASA's Marshall Space Flight Center (Fig. 6.21) using 3D printing technology. The scientists used different methods to create lab-grown human liver tissue that was robust enough to survive and function in a manner similar to that inside the human body. Using different 3D printing technologies, they built a cubic tissue about 1 cm thick and were able to keep it running in the lab for 30 days. Human tissues rely on blood vessels to provide nutrients and oxygen to cells and remove metabolic waste, a process known as perfusion. When perfusion tests were performed on the lab-grown human tissues, no leakage occurred when fluid passed through the tissue. This research could help achieve the growth and long-term survival of 3D tissues for research and treatment, and ultimately organ packaging and replacement.

6.3.3 Emergency Bioprinter

The emergency bioprinter in Fig. 6.22 is manually operated and consists only of a handle, dispenser, print head, guide wheel, and two bio-ink cartridges. This technology does not use real human cells, but fluorescent particles. When combined with two fast-curing gels, these particles produce a plaster-like wound covering that will

Fig. 6.21 NASA's Vascular Tissue Challenge

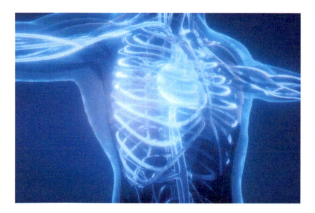

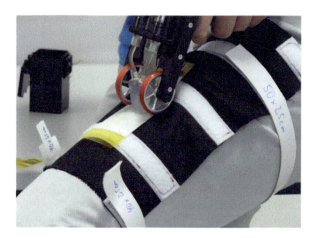

Fig. 6.22 Handheld emergency bioprinter

be printed on the astronaut for emergency use. The astronaut is then sent back to Earth for further examination and treatment.

6.3.4 3D Printing Ensures Astronaut Life Systems

In the future, astronauts venturing into deep space may use 3D printed skin, bones, and even entire organs for treatment (Fig. 6.23). Apollo astronauts once spent about 12 days for a trip to the moon. During the moon landing, a small medical bag was carried, containing bandages, antibiotics, and aspirin. In the future, space travelers who are months or years away from Earth will need more flexible medical support.

Just as standard 3D printers use plastic or metal to build three-dimensional objects, bio-3D printers use "bio-ink" based on human cells, as well as the nutrients and materials needed to regenerate body tissues (such as skin, bones, and cartilage),

6.3 3D Printing of Biomimetic Materials

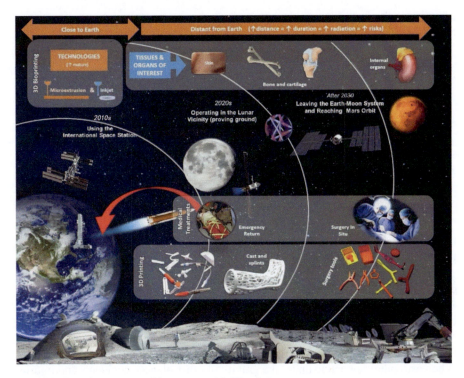

Fig. 6.23 Ensuring astronaut life systems with 3D printing

to provide stability and optimal physical conditions, promoting the reconstruction of body tissues (Figs. 6.24 and 6.25).

Fig. 6.24 3D printed artery

Fig. 6.25 The process of 3D printing bones

However, printing entire organs will be a more challenging task, which is perhaps the goal for the next decade, involving precise combinations of various cells and tissue types, as well as overall coordinated work.

Compared to today's low-Earth orbit space activities, long-distance missions to distant destinations will face very different challenges. In the event of a medical emergency, it is impossible to return to the ground quickly, and astronauts must be treated on the spot.

As astronauts having bone and muscle weakening problems in a weightless state, it may become more severe when astronauts stay in orbit for years rather than months. When astronauts are active under the gravity of other planets, accidents and injuries may also occur, increasing the risk of bone and joint injuries as well as skin injuries caused by slips and accidents, which even exceeds the risk brought by radiation exposure.

Due to the limited interior space in the spacecraft, it is impossible to prepare for all possibilities, and it is difficult to provide complete medical materials and conditions. However, bio-3D printing capabilities will allow for more flexible and universal responses to emergencies. For example, severe burns are usually treated with skin grafts from other parts of the patient's body, which involves secondary damage to the transplant area. Conversely, if new skin could be grown and bioprinted from the patient's own cells, and then directly transplanted, or if damaged parts of the body, including skin, bones, or internal organs, could be replaced directly with printed tissues in custom shapes, it not only provides new options for real-time treatment during space exploration, but also opens up new avenues for treatment on Earth.

Frankly speaking, there are so many problems to be solved in order to achieve this purpose. An unsolved problem is that how bio-printed structures mature after printing, and how their implementation in the human body could be affected by changes in space conditions. In addition, the surgery required for transplanting tissues also needs to be reconsidered. The sterile environment, equipment, and trained personnel of ground operating rooms, as well as easily provided disposable surgical supplies, will become difficult in space. Therefore, surgical robots may be needed in space to help fill in the gaps in surgical skills, operating autonomously through artificial intelligence, rather than being remotely controlled by earth-based surgeons,

6.4 Status of Printing Materials

as the communication delays involved in deep space missions make direct remote medical treatment impractical.

6.4 Status of Printing Materials

6.4.1 Heat Resistance of 3D Printed Metal

Future lunar landers (Fig. 6.26) may be equipped with 3D printed rocket engine parts, helping to reduce overall manufacturing costs and shorten production time. For rocket engines, the heat resistance of printed parts is a crucial characteristic. High-strength iron-nickel high-temperature alloy nozzles are printed using a method called laser powder directed energy deposition. Through a series of hot fire tests (Fig. 6.27), NASA has verified that two types of 3D printed engine parts (a copper alloy combustion chamber and a nozzle made of high-strength hydrogen-resistant alloy) can withstand the same extreme combustion environment as traditionally manufactured metal structures. These hot fire tests are a key step for these parts to be used in future lunar and Mars missions. Throughout the testing process, engineers collected a large amount of data, including pressure and temperature measurements in the coolant channels and main chamber, as well as high-speed and high-resolution videos of the exhaust plume and chamber throat. The research team also calculated the performance of the chamber and the overall efficiency of the engine using propellant. These tests aim to enable these 3D printed rocket engine parts and other 3D printed hardware to be used in future lunar landers.

Fig. 6.26 Future lunar lander and conceptual lunar base

Fig. 6.27 Hot fire tests of 3D printed rocket engine parts

6.4.2 Laser 3D Printing X-ray Imaging

To further understand the potential laser-material interaction mechanisms involved in laser-based metal 3D printing, the printing process can be observed through high-speed synchrotron radiation X-ray imaging. This is also a study supported by the European AMAZE project (Fig. 6.28). When the laser beam contacts the powder particles, they fuse into individual solid beads through laser melting, on a timescale of microseconds. This study reveals the basic physical principles in the melt track deposition process, showing that laser-induced gas/vapor jets promote the formation of melt tracks and spallation areas through splashing.

6.4.3 Surface Condition of 3D Printing

Figure 6.29 is a scanning electron microscope close-up of the surface of a part produced by 3D printing. The image shows a cross-section of an internal cavity. The material is Ti_6Al_4V, treated by selective laser melting, and then post-processed to remove loose powder in its cavity. The part is cut open to expose the cavity. The observation results show that significant improvements need to be made to the cleaning process to ensure a surface free of loose particles. Due to its printing manufacturing process, loose particles will adhere to the solid surface, which could cause problems when used in space. The largest particle diameter observed in Fig. 6.29 is about 100 μm, equivalent to the thickness of a human hair.

6.4 Status of Printing Materials

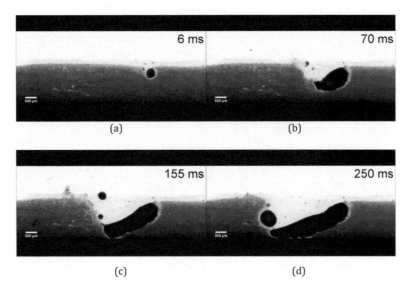

Fig. 6.28 Close-up of laser 3D printing X-ray imaging at different time, (a) 6 ms, (b) 70 ms, (c) 155 ms, (d) 250 ms

Fig. 6.29 Electron microscope image of the surface morphology of a 3D printed part

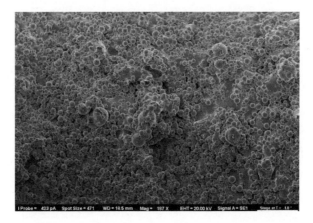

Chapter 7
Technical Applications of Space 3D Printing

7.1 High-Temperature Resistant 3D Printing Technology

7.1.1 3D Printing of High-Temperature Resistant Nozzles

Metals with complex heat dissipation structures made using 3D printing can withstand high temperatures. NASA's Rapid Analysis and Manufacturing Propulsion Technology (RAMPT) project is an additive manufacturing technology, using metal powder and laser 3D printing rocket engine parts (Fig. 7.1). This method, known as blown powder directed energy deposition, reduces the cost of producing large complex engine parts (such as nozzles and combustion chambers) and shortens delivery times. The printing method injects metal powder into a molten metal pool heated by a laser (also known as a melt pool), the nozzle that blows out the powder and the laser optics are integrated into a print head, and this print head is connected to a robotic arm. The print head moves according to a computer-determined pattern, building one layer at a time. This manufacturing method has many advantages, including the ability to produce very large products. It can also be used to print very complex parts, including engine nozzles with internal coolant channels. Rocket engine nozzles with internal coolant channels run low-temperature propellant through these channels to help the nozzle maintain a safe temperature. Manufacturing nozzles in a traditional way is a challenging process that may take a long time. Blown powder directed energy deposition additive manufacturing can create large-scale parts with complex internal features, which was difficult to achieve before. It can greatly reduce the time and cost of manufacturing cooling nozzles with internal channels and other key parts of rockets.

The RAMPT team recently used this technology to produce one of NASA's largest printed nozzles, with a diameter of 40 inches (101.6 cm) and a height of 38 inches (96.52 cm), with integrated cooling channels (Fig. 7.2). The manufacture of this nozzle took only 30 days, while the traditional welding method would take nearly a year. Using this new additive manufacturing technology to produce channel wall

Fig. 7.1 Rocket engine parts printed with metal powder and laser 3D by NASA's Rapid Analysis and Manufacturing Propulsion Technology (RAMPT) project

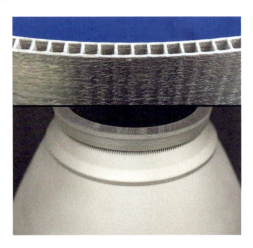

nozzles and other parts will help the spacecraft R&D team manufacture SLS engines at the required scale, shorten the schedule and reduce costs. In a series of rigorous hot fire tests, engineers will subject the nozzle to the combustion temperature of 6000 ℃ and the continuous pressure faced during launch to prove the durability and performance of products produced by this new directed energy deposition technology.

In addition, a team of engineers at NASA's Marshall Space Flight Center has developed and advanced a new process called Laser Wire Direct Closeout (LWDC) to significantly reduce the time to manufacture lower-cost nozzles. LWDC is a process different from most 3D printing technologies (based on powder and layer-by-layer manufacturing), it uses a free-form directed energy line deposition process. This new patented technology has the potential to reduce construction time from several months to a few weeks. The LWDC method uses a metal wire-based additive manufacturing

Fig. 7.2 Overall configuration of the 3D printed rocket engine nozzle

Fig. 7.3 Nozzle hot fire test

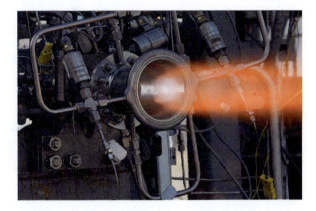

process to precisely close the nozzle coolant channels, which contain high-pressure coolant liquid, to protect the nozzle walls that must withstand high temperatures.

The nozzle is actively cooled, which means that the propellant used in the combustion cycle passes through the nozzle (Fig. 7.3), cooling the nozzle wall to some extent so that it does not overheat. To fabricate this actively cooled nozzle, a series of channels need to be made inside the nozzle, but they must then be closed and sealed to accommodate high-pressure coolant. The new patented technology closes the coolant channel and prints in place to form a support sleeve that responds to structural loads during engine operation.

7.1.2 3D Printed Platinum Alloy Thruster Chamber

A project by the European Space Agency called Advanced Satellite Thrust Chamber Additive Manufacturing Technology (AMTAC) printed the combustion chamber and nozzle of a 10 N thruster in platinum alloy (Fig. 7.4) using a laser beam applied to a metal powder bed. A hot fire test was then conducted, reaching a maximum temperature of 1253 ℃. The production and testing of the thruster prototype were completed at the Airbus Defence and Space facility in Lampoldshausen, Germany.

7.1.3 3D Printed Insulation Technology

Space launch systems, when launching spacecraft and other cargo to lunar orbit, will face harsh conditions and extreme temperatures in flight, requiring robust temperature protection. Technicians and engineers use 3D printing technology to apply thermal protection systems to smaller, more complex parts of the rocket.

By using spray foam or traditional insulation materials, large and small parts of the rocket are protected from heat during launch and the low temperature of the propellant

Fig. 7.4 3D printed platinum alloy nozzle

in large tanks is maintained. However, small hardware or narrow areas, such as the internal pipes of the engine part, require technicians to manually spray foam or use foam casting, and in some cases, use 3D printed molds. The foam that is mixed and poured into the mold will expand and perfectly fit the part. It reduces the complex and tedious post-processing work, thereby shortening the overall processing procedure. NASA and Boeing engineers conducted a large amount of technical development and identification of foam casting tests in the early stages of the project. Using these data, the team developed a series of processes that reduced the time required for a single 3D printed mold. It allows more time to focus on tackling the key technical requirements that each foam must meet. It simplifies the process from 3D printing to casting application.

7.2 Complex Structure 3D Printing Technology

7.2.1 3D Printed High Complexity Rocket Turbo-Pump

The turbo-pump is one of the most complex 3D printed rocket engine parts ever. In a series of successful tests conducted at NASA's Marshall Space Flight Center using liquid hydrogen propellant, the turbo-pump achieved a speed of over 90,000 revolutions per minute, as shown in Fig. 7.5. This turbo-pump is manufactured by 3D printing, and the parts are 45% less than similar pumps made by traditional manufacturing.

3D printing is a key technology for enhancing spacecraft design. The turbo-pump is a key component of the rocket engine, its turbine rotates and generates over 2000 horsepower (1491 kW) of power (twice the power of a NASCAR engine). In 15 tests, the turbo-pump reached full power, delivering 1200 gal (4543 L) of low-temperature liquid hydrogen per minute. The final stage rocket engine provides power with a thrust of 156 kN. The fuel pump and its components are designed and manufactured using 3D printing. The design is loaded into the 3D printer as a running program, which then manufactures each component by layering metal

7.2 Complex Structure 3D Printing Technology

Fig. 7.5 Turbine machine with a speed exceeding 90,000 r/min

powder and melting it with a laser. During testing, the 3D printed turbo-pump is exposed to the extreme environment inside the rocket engine, with fuel burning at temperatures exceeding 6000 ℉ (3315 ℃) to generate thrust. The turbo-pump delivers fuel in the form of liquid hydrogen cooled to below −400 ℉ (−240 ℃). The tests help ensure that the 3D printed parts operate successfully under these harsh conditions.

7.2.2 3D Printing of Complex Aircraft Icing Shapes

Aerospace R&D personnel are using the most modern research tools (including 3D printing) to generate new research data (Fig. 7.6), which will help aircraft manufacturers and operators more effectively deal with one of the important safety challenges of aerospaceicing.

The icing problem has been studied in the space industry since before World War II. The new set of data allows researchers to better understand the formation of ice and its impact on aircraft, with the key factor being the use of 3D printing as a research tool. Historically, icing research has relied on producing real ice in specially equipped wind tunnels, blowing super-cooled droplets onto the aircraft surface (usually the wings), and then freezing upon contact.

Although the shapes of ice produced under these controlled conditions are very consistent with those formed in nature, the methods used to record and analyze these shapes are relatively simple. The most common method is to cut a mouth on the ice with a heated metal plate, insert a piece of cardboard, and then trace the outline of the ice with a pencil. Measurements of these outlines provide basic data for computer calculations, running simulations to understand and predict the impact of various ice shapes on aircraft aerodynamics. However, due to the rough characteristics of these drawings, the details of the ice shape are lost, meaning that the resulting computer

Fig. 7.6 Using a 3D laser scanner to measure the icing of large wing sections

codes cannot fully represent the real situation in the air. Over the years, researchers have tried to manually create and measure more complex shapes using traditional model-making methods, including using molds and castings. Then, these models will be connected to the aircraft surface and tested in the wind tunnel.

With the emergence of advanced 3D printing technology, and the ability to create and manipulate 3D models of scanned objects in computers, icing research has entered a new era. Figure 7.7 is an example of a complex ice shape printed by a 3D printer, which NASA's Glenn Research Center uses for studying the impact of icing on aircraft aerodynamics.

Fig. 7.7 Complex ice shapes printed by a 3D printer

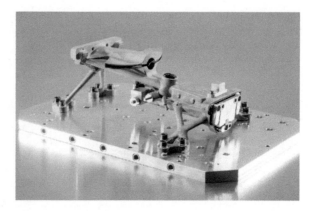

Fig. 7.8 Gas identification telescope manufactured using 3D printing

7.2.3 3D Printing of High-Precision Gas Identification Telescopes

Figure 7.8 is a space telescope designed by a Dutch team from the European Space Agency, used to identify ozone and other trace gases in the Earth's atmosphere. It consists of three main parts, including two mirrors of the telescope, a total of nine parts, printed with flight-grade aluminum alloy. Compared to the original 2.8 kg version (NASA's EOS-Aura Ozone Monitoring Instrument telescope in operation), this version weighs only 0.76 kg, about 73% lighter, and the measurement accuracy has not decreased.

The instrument requires extremely precise optical alignment to do its job, identifying the spectral fingerprints of trace gases in the atmosphere. The optical-mechanical structure places high demands on materials, production processes, and the design team.

The advantage of 3D printing allows this instrument to complete tasks with few parts. Because these parts are built with unprecedented complex geometric shapes, they can combine functions together. It saves valuable weight and reduces the workload of designing, assembling, and testing systems. At the same time, it could achieve all the necessary performance of a robust top-level optical instrument. The printing process also means that the complex internal structure of the telescope can be completed at once, including multiple baffles and fixed two-sided mirrors used to reduce unnecessary internal stray light.

7.2.4 3D Printing of a Complete Storable Thrust Chamber

The European Space Agency further demonstrated the maturity of additive manufacturing technology in rocket engine design for rocket stages, on-orbit transportation,

Fig. 7.9 3D printed full-scale thrust chamber

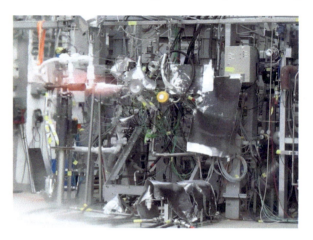

micro launchers, and exploration spacecraft (such as lunar landers and lunar ascent vehicles) by 3D printing a full-scale rocket thrust chamber.

The thrust chamber, entirely manufactured by 3D printing (Fig. 7.9), is designed for storable propellants, which can be stored in liquid form at room temperature, making the rocket engine more reliable during months-long flight missions and easier to reignite. The 3D printed combustion chamber has a standard thrust of 2.5 kN and was fired for 560 s in the Lampoldshausen test center of the German Aerospace Center.

The combustion chamber, developed by the Ariane Group in the European Space Agency's Future Launchers Preparatory Program, helps to study the flow and heat transfer phenomena on the surface of parts created by 3D printing. A complex cooling channel needs to be arranged in the combustion chamber structure to cool the chamber walls.

For hot launches and actual flight processes, 3D printed parts present a series of challenges, especially when dealing with fine, complex structures, such as cooling channels. The hot fire test proved that the 3D printing process is an effective manufacturing method, and also helps to promote understanding of the flow phenomena inside the additive manufacturing rocket engine. With 3D printing technology, researchers can more flexibly separate the cooling system from the combustion process to study the thermodynamic and fluid dynamic characteristics of additive manufacturing structures and surface characteristics.

7.3 High Utilization of 3D Printing Technology

7.3.1 3D Printing Aids in the Search for Black Holes

A X-ray telescope developed by European Space Agency aimed at searching for supermassive black holes (Fig. 7.10) uses a novel 3D printing technology called plasma metal deposition to build. The space telescope (Athena) is scheduled to launch in 2033. Plasma metal deposition technology is also a candidate for manufacturing large space parts in the future, such as Athena's optical bench (Fig. 7.11, which is responsible for aligning and fixing about 600 mirror modules). It will be the largest part ever printed with titanium alloy. The overall shape has a diameter of about 3 m, and the precision must be controlled within tens of micrometers.

The entire process chain of this optical bench uses titanium alloy as the metal powder or wire material for printing. Tests show that the part has good mechanical

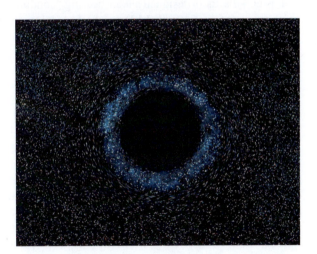

Fig. 7.10 Cosmic black hole

Fig. 7.11 Athena's optical bench after 3D printing and milling. *Source* ESA

properties and precision machining performance, marking that high-precision tasks be completed successfully using 3D printing technology. Comparatively, if traditional manufacturing methods are used, i.e., milling from a block, it will result in more than 80% material waste. By using the plasma metal deposition technology, material and cost can be greatly saved.

7.3.2 Building a Lunar Base with 3D Printing

Using local materials selected by a 3D printer can make the establishment of the lunar base simpler (Fig. 7.12). Currently, relevant aerospace agencies begun testing the feasibility of using lunar soil for 3D printing. Now, 3D printing technology on Earth can produce the entire framework structure of a building, but it still needs to be confirmed whether this can also be used to build a lunar base.

Foster + Partners designed a load-bearing "conduit" dome design, using a honeycomb structure wall (Fig. 7.13) to shield micrometeoroids and space radiation, and combined with a pressurized inflatable device to ensure astronaut activities. The hollow closed-cell structure, similar to bird bones, can provide a solution that balances structural strength and weight.

The design of the base was based on the characteristics of 3D printed lunar soil and a 1.5 t building block sample was made. This 3D printing method provides a potential technical means for promoting lunar settlement (Fig. 7.14), which can reduce logistics costs from Earth.

The UK's Monolite company provides the D-Shape printer, which has a moving print nozzle array on a 6 m long frame, spraying adhesive solution onto sand-like building materials. The wall printed with this printer is shown in Fig. 7.15.

During the printing process, the artificial lunar material needs to be mixed with magnesium oxide first, turning it into "paper" that can be printed, and then used as structural "ink". The company uses a combination to transform the material into a stone-like solid. Currently, Monolite's D-Shape printer works at a speed of about

Fig. 7.12 Concept of 3D printed lunar base

7.3 High Utilization of 3D Printing Technology

Fig. 7.13 Large hollow building block

Fig. 7.14 Multi-dome base in design concept

Fig. 7.15 Wall printed with the D-Shape printer from Monolite, UK

Fig. 7.16 Sculpture molded by Alta company using 3D printing

2 m/h, and the next-generation design printer should reach 3.5 m/h, and it is expected to complete the construction of the entire building within a week.

In addition, the Italian space research company Alta (Fig. 7.16) cooperates with the University of Pisa to adapt 3D printing technology to lunar missions, ensure process quality control, and also assess its impact on working in a vacuum. Alta's printing process is based on the application of liquid, considering that unprotected liquid will boil in a vacuum, so the 3D printer's nozzle needs to be inserted below the plaster layer. The study found that 2 mm small droplets are retained in the soil by capillary forces, which means that the printing process can indeed be completed in a vacuum.

Another factor to consider is that 3D printing works best at room temperature, but in most areas of the moon, the temperature varies greatly over the course of two weeks of day and night. For future lunar settlements, the lunar poles can provide the most moderate temperature range (Fig. 7.17).

7.3.3 3D Printing Helps Fight "COVID-19"

Thanks to the versatility and customization of 3D printing, 3D printers at the European Astronaut Centre (EAC) of the European Space Agency have been put to work in the fight against the novel coronavirus pneumonia (referred to as "COVID-19"). Two open-source 3D printers are used to produce face mask components, and the completed masks will be delivered to hospitals in need (Figs. 7.18 and 7.19).

7.3 High Utilization of 3D Printing Technology

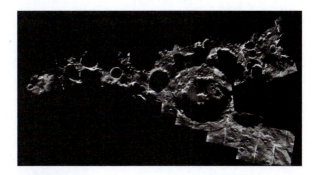

Fig. 7.17 Satellite image of the lunar south pole

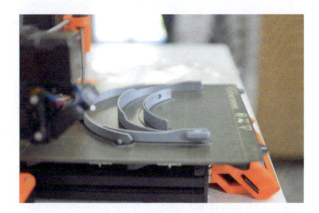

Fig. 7.18 3D printed mask bracket

Fig. 7.19 Face shield parts 3D printed by EAC

Following the same standardized mask printing design, the procedure has been optimized through ergonomics to ensure the rapid, efficient, and consistent production of mask headbands and brackets, contributing timely to the final product. Each mask consists of four parts: a 3D printed headband and a 3D printed bracket, both

Fig. 7.20 Assembled mask (ready-to-use, with filter mask)

of which are installed in a mask made of a transparent plastic sheet, and the fourth part is an elastic band that allows the mask to be comfortably secured to a person's head (Fig. 7.20).

EAC is responsible for printing the headbands and brackets. It takes about 1.5–3 h to print all the parts for each set.

3D printing has become a valuable tool for future space travel because it brings production closer to the point of use and enables astronauts to produce parts when they need them, rather than carrying a full set of spare parts. It also allows discarded materials to be recycled into usable items and could potentially be used to build lunar structures.

7.3.4 3D Printing of Astronaut Medical Tools

As a technology that is expected to become universal, 3D printing can provide customized auxiliary tools for checking the physical condition of astronauts. NASA hopes to develop 3D printable designs for instruments on the International Space Station (ISS) that can handle liquids like blood samples without spilling under microgravity. These tools allow astronauts to analyze biological samples without having to send them back to Earth.

Once astronauts enter space, they must perform all analyses themselves. Therefore, there is a need to develop an automatic system to study molecular biology with minimal human intervention. One of the challenges of preparing samples is handling liquids under microgravity. Astronauts collect various samples, including their own saliva and blood, as well as microbes swabbed from the walls of the International Space Station. These samples are then mixed with water so they can be injected into

7.4 Integrated 3D Printing Technology

Fig. 7.21 The key part of the extractor is a 3D printed plastic box

instruments for analysis. Without the proper tools, samples may spill, float, or form bubbles, affecting the test results.

In 2016, NASA first sequenced DNA in space. Astronauts used the MinION handheld sequencing tool and developed an automatic DNA/RNA extractor based on it for MinION. The device prepares the sample. The key part of this extractor is a 3D printed plastic box (Fig. 7.21), used to extract nucleic acids from the sample for MinION sequencing.

7.4 Integrated 3D Printing Technology

7.4.1 3D Printed Miniature Sensors

The core of this work on 3D printed miniaturized sensors, funded by NASA's Space Technology Mission Directorate (STMD) Early Career Initiative (ECI), is a 3D printing system developed by Ahmed Busnina and a research team at Northeastern University in Boston. This 3D printing system is similar to a printer that produces currency or newspapers. However, instead of using ink, this printer applies nanomaterials layer by layer onto a substrate to create miniaturized sensors (Fig. 7.22). Each resulting sensor can detect different gases, pressure levels, or temperatures.

Nanomaterials, such as carbon nanotubes, graphene, molybdenum disulfide, and other materials, exhibit interesting physical properties. They have high sensitivity and stability under extreme conditions. This kind of materials is lightweight, which exhibits strong resistance to radiation and requires less power energy loss. It is an ideal choice for space applications. The Sultana team designed a sensor platform, and Northeastern University in the United States used its nano-level pad printing system to utilize nanomaterials. After printing, the Sultana team functionalized individual sensors by depositing additional layers of nanoparticles to enhance their sensitivity, integrated the sensors with readout electronics, and packaged the entire platform. This kind of approach is very different from the current way of manufacturing multi-functional sensor platforms. 3D printing technology allows technicians to print a

Fig. 7.22 NASA miniaturized sensors

set of sensors on one platform, rather than manufacturing one sensor at a time and then integrating it with other components, greatly simplifying the integration and packaging process.

The advantage of this technology is that all sensors and some circuits can be printed on the same substrate, which can be rigid or flexible, eliminating many packaging and integration problems. All sensors can be integrated on the same chip, printed layer by layer. The Sultana team used the same technology to manufacture and demonstrate single sensors made of materials such as carbon nanotubes and molybdenum disulfide. The prepared sensors are very sensitive, with an accuracy of one in a million.

7.4.2 3D Printed Miniature Satellite Body

The 3D printed Micro-Satellite (Fig. 7.23) is made of Polyether Ether Ketone (PEEK). By incorporating conductive particles, the electrical circuits can be embedded in these Micro-Satellites. In the future, when payload instruments, circuit boards, and solar panels are installed, this type of miniature satellite can be put into use at any time.

7.5 Lightweight 3D Printing Technology

The European Space Agency has used 3D printing technology to manufacture an antenna support pillar, reducing its weight by 46%; an internal silver-coated RF filter, reducing its weight by 50%, and shortening its manufacturing time by several

7.5 Lightweight 3D Printing Technology

Fig. 7.23 Integrated miniature satellite body using 3D printing. *Source* ESA

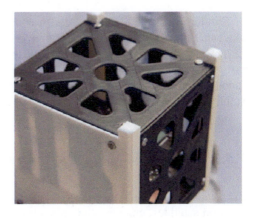

weeks. These all show that 3D printing is an important means for aerospace products to achieve lightweight.

7.5.1 Lightweight 3D Printed Components

Adopting a honeycomb lattice (Fig. 7.24) can effectively reduce the weight and cost of the rocket thrust chamber and nozzle, and has significant potential for reducing total weight of the satellite. Considering that the engine temperature can reach up to 2500 ℃, it is also necessary to improve its thermal elasticity. Compared to standard solid objects, the surface area of this lattice is greatly increased, and it can enhance radiative cooling.

Fig. 7.24 Lattice structure of a 3D printed engine nozzle

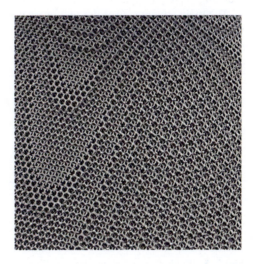

Similar 3D printed lattices are also considered for use as more durable propellant catalyst beds to avoid the performance degradation of standard particle catalysts during the lifespan of the propulsion system. In addition, researchers are considering using this specific 3D printed lattice design for propellant management devices (sponge-like structures inside the propellant tank) to prevent bubbles and ensure stable propellant performance.

7.5.2 3D Printed Metal Microwave Devices

3D printing using metal material is expected to improve the performance of communication satellites, enabling the production of radio frequency (RF) filters or other components at a lower cost.

RF filters are essential parts in space communication system. When RF signals transmit in space, these filters work like gatekeepers, filtering out unwanted frequencies while allowing selected channels to pass through. A typical modern communication satellite may carry hundreds of such filters (Fig. 7.25). Their complex internal contours are specifically designed for certain frequencies to allow the use of multiple signal beams.

Building on previous research by the European Space Agency, Airbus Defence and Space in the UK developed a 3D printed RF filter prototype (Fig. 7.26), which has proven its space suitability through rigorous testing. Traditional metal waveguide RF filters are cut into two sections and then fixed together—this is usually designed for feasibility rather than performance. In contrast, 3D printing produces a single integral part. The new technology brings new design freedom, reduces restrictions, and differs greatly from traditional methods, reducing assembly processes while also reducing spare parts, achieving product lightweighting.

The main advantages of the one-piece design achieved by 3D printing are lighter weight, lower cost, and shorter production time. The weight is reduced because screws are greatly reduced for easier assembly. In addition, direct metal printing has

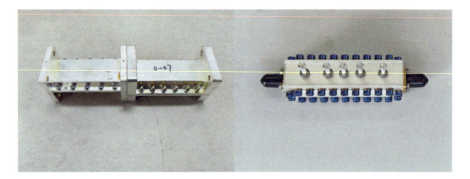

Fig. 7.25 RF filters in communication satellites

7.5 Lightweight 3D Printing Technology

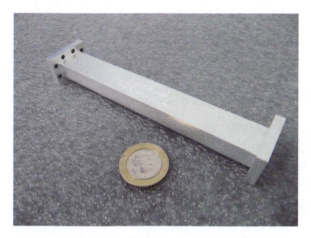

Fig. 7.26 3D printed radio frequency (RF) filter prototype. *Source* ESA

another advantage: the outer contour is closer to the inner contour, so less metal is used to form the cavity wall. The advantages in cost and time come from the reduction in assembly and post-processing work.

Since the printing process uses metal powder particles and melts them together with a laser, the challenge is how to ultimately obtain a sufficiently smooth finished surface. The microscopic topography of 3D printed parts is different from that of machined parts. Machined surfaces are often relatively flat, and the roughness depends on the accuracy of post-processing, while the surface accuracy of printing is related to factors such as the size of powder particles, melting adhesion, and porosity. For the internal part of the integrated filters, it is hard to smoothen the surface through post-processing. The design and production process of 3D printed microwave parts is shown in Fig. 7.27.

Three aluminum parts were printed using different post-processing methods, and then tested under simulated launch and orbital conditions, including vibration, vacuum, and extreme temperatures. All three parts met or exceeded the required performance, with the best performing being a filter silver-plated by electrolysis. The tests showed that this process can be used for future space microwave payload production, demonstrating faster turnaround times, reduced production costs, and a 50% reduction in weight. Currently, researchers are trying to further reduce weight through design optimization, including integrating different functions and components in subsequent RF filter designs (Fig. 7.28).

To further optimize waveguide RF components, a multidisciplinary team is needed. Material engineers, microwave engineers, thermal and mechanical engineers, along with 3D printing researchers, are using next-generation computer-aided design tools to pioneer new methods for 3D printing.

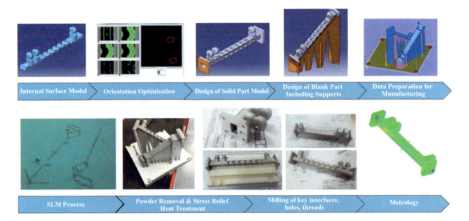

Fig. 7.27 Design and production process of 3D printed microwave parts

Fig. 7.28 Assembled RF filter

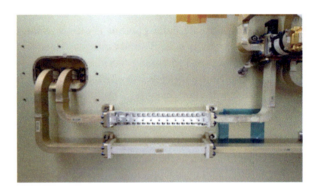

7.5.3 Integrated 3D Printed Satellite Bracket

The integrated 3D printed lattice sensor bracket prototype for satellites shown in Fig. 7.29 was designed by LKE, a Czech company, and manufactured by Brno University of Technology. Produced by selective laser melting, the one-piece metal lattice part has a reduced weight compared to standard parts. It weighs only 164 g, while the weight of a similar product produced traditionally is 222 g.

7.5 Lightweight 3D Printing Technology

Fig. 7.29 Integrated 3D printed lattice sensor bracket prototype

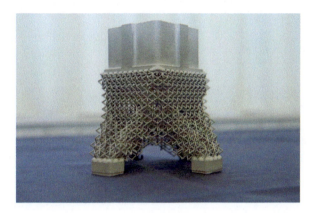

Chapter 8
Future Development of 3D Printing in Space

Building bases on the moon, Mars, or even further planets is gradually laying the foundation for humans to venture into space. With this as a fulcrum, humans begin to imagine journeys to even more distant planets. In this process, what role will 3D printing technology play?

8.1 3D Printing of Food in Space

To complete long-term space travel, food and air supplies for astronauts are essential and crucial.

Currently, for food supply on the space station, individually packaged, long-lasting items are mainly used. Astronauts can choose the food they want to carry in advance, then launch it into space all at once, or rely on subsequent space cargo ships for resupply.

How to achieve a diverse food supply and comprehensive nutrition supplement during long-term voyages, especially to solve the problem of food preservation, is a fundamental problem that astronauts need to solve when staying in space for a long time. The current space food supply system can no longer meet the nutritional needs and shelf-life requirements needed for Mars exploration missions or other long-term tasks.

3D printing provides a broader implementation plan for food preparation in space. By carrying long-lasting raw materials, customized food production can be achieved to meet the safety, acceptability, diversity, and nutritional stability requirements of long-term exploration missions, while consuming the least spacecraft resources and crew time.

Currently, research on nutrient pack production is being carried out on the International Space Station (Fig. 8.1). Using microorganisms such as bread yeast, basic

Fig. 8.1 The production process of nutrient packs in the International Space Station

materials such as dry powder and water are used to provide astronauts with fresh nutrients. By completing the microbial growth process on the space station, the produced products are then frozen and sent back to Earth for further analysis, to study whether fresh nutrients necessary for human survival in space can be produced.

8.2 Resupply of Spaceship Parts During Interstellar Travel

When people travel interstellarly in spaceships, or from Earth to other planets, they often encounter the failure of spaceship parts, which requires resupply. It is a common scene in many science fiction movies. In the real world, this is a problem that space scientists need to consider from the beginning. Unlike the Earth's space station that can rely on a limited number of spaceship round trips or rocket launches for resource resupply, in interstellar travel, it is almost impossible to get stable supplies from Earth, highlighting the importance of 3D printing technology once again.

For the resupply of spaceship parts during interstellar travel, there are mainly two conceived methods. One is to disassemble existing devices to obtain raw materials, analyze faulty components, and then print faulty components (such as connectors, capacitors, inductors, resistors, etc.) mainly for repair. The other is to print new components or tools. Whether fault repair, component resupply, or re-customization, raw materials are indispensable. Recycling existing materials for printing is a good choice.

Figure 8.2 is a developed electronic component 3D printer. Designer Jones Glenn traced the inspiration for his electronic component 3D printer back to "Star Trek". In the late 1970s, he was a high school student watching "Star Trek", and was shocked by the "replicator" shown in it. This is a machine exhibited at the exhibition, which can produce food, clothing, ship parts, and anything else needed as required.

Glenn submitted an application for the project to NASA's Small Business Technology Transfer (STTR) program, which funds R&D projects that contribute to

8.2 Resupply of Spaceship Parts During Interstellar Travel

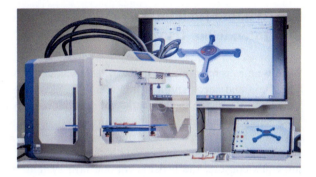

Fig. 8.2 Electronic component 3D printer *Source* NASA

NASA missions. NASA approved Glenn's project and provided about $1 million in support during the first and second phases of the STTR program. He also received a Phase III Small Business Innovation Research contract.

Because the ability to 3D print parts in space can support long-term space missions, NASA has been studying this technology for many years. Especially recently, electronic technology has become a research focus because these high-precision parts and equipment have failed in the past on the International Space Station.

For the development of 3D printers for electronic components, it is necessary to develop printable materials with the required electrical characteristics and corresponding 3D printers. Since the production of electronic components usually requires at least several different materials, the printer needs to be able to print multiple materials without manually changing the print head.

Currently, eForge uses a 3D printing process called Fused Deposition Modeling (FDM), which first heats the material to a liquid state, then sprays it out through a nozzle, similar to hot glue sprayed from a glue gun, and leaves it in space to cool and harden. Glenn and his team have developed six basic materials for the initial operation of the eForge printer and have applied for patents. These include conductive materials, insulating materials, resistors, and capacitor filaments, in addition to a new type of semiconductor material. According to the company, eForge's semiconductor materials can be used to manufacture switches, communication equipment, and solar cells, as well as diodes and transistors for integrated circuits, computers, and amplifiers.

In addition to being used in space, this type of printer can also be used in maker spaces in universities and high schools. Students can create their own modelled and tested circuits and devices, and design, print, and test them to ensure they work. If not, they can be redesigned and reprinted within a few minutes.

8.3 What to Do if an Astronaut Gets Seriously Ill During Interstellar Travel

What if an astronaut or interstellar traveler suddenly becomes seriously ill during a long flight and needs surgical treatment? Researchers from various countries have preliminarily realized the possibility of on-orbit remote surgery by combining the da Vinci surgical robot and satellite communication technology. The da Vinci Surgical System is an advanced robotic platform designed to perform complex surgeries in a minimally invasive way. The da Vinci surgical robot consists of three parts: the surgeon's console, the bedside robotic arm system, and the imaging system. For on-orbit remote da Vinci surgery, accurate, real-time communication provided by satellites is needed to help doctors on the ground understand the astronaut's physical condition and health status in a timely manner and make accurate surgical judgments.

And the human organs needed for surgery would require the help of 3D printing technology.

Although it sounds horrifying, using cells as the raw material for 3D printing, printing human tissues and even organs is indeed not new, and has become an important interdisciplinary field combining 3D printing technology and biological science.

Currently, the use of 3D printing technology to obtain stable and reliable human tissues and even organs to deal with sudden changes during long space travels has become a frontier technology that scientists have begun to explore, and some research progress has been made in the early stage.

In 2016, NASA launched a space competition event named "Vascular Tissue Challenge", as part of the Centennial Challenge of the NASA Marshall Space Flight Center. In 2021, two groups of scientists from the Wake Forest Institute for Regenerative Medicine (WFIRM) in Winston-Salem, North Carolina, won first and second place in this competition. These two groups of scientists, representing the Winston team and the WFIRM team, used different methods to create human liver tissues that can grow in the lab. These tissues are robust enough to survive and function in a manner similar to that inside the human body. Each team used different 3D printing technologies to construct a cubic tissue about 1 cm thick that could work in the lab for 30 days.

Figures 8.3 and 8.4 show the work of the Winston team, the first-place winner of the Vascular Tissue Challenge. They used a cavity to store the printed organ tissue and conducted a perfusion test on it. Human tissues rely on blood vessels to provide nutrients and oxygen to cells and remove metabolic waste, a process known as perfusion. Because this process is difficult to replicate in engineered tissues, NASA required the participating teams to develop tissues with artificial blood vessels and provide a plan for conducting perfusion tests in them.

According to the rules of the challenge, the winning team used 3D printing technology to create a gel-like mold or scaffold and designed a channel network to maintain sufficient oxygen and nutrient levels, keeping the printed tissue alive during the 30-day trial. The value of artificial tissues depends entirely on their degree of

8.3 What to Do if an Astronaut Gets Seriously Ill During Interstellar Travel

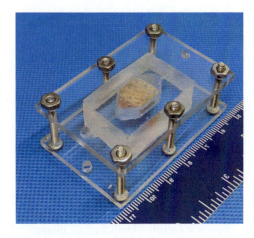

Fig. 8.3 Winston team's work for the Vascular Tissue Challenge

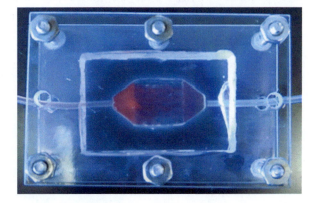

Fig. 8.4 When the Winston team's work for the Vascular Tissue Challenge underwent a perfusion test, the liquid passed through the tissue without leaking

imitation accuracy in replicating real human conditions. These requirements are very precise and vary for different organs, making the task extremely complex.

The Winston team and the WFIRM team used different 3D printing designs and different materials to create living tissues containing cell types found in the human liver.

For long-term space travel, 3D printing of human organs is showing promise, providing a possible solution to the crisis of organ shortage for future astronauts, and also providing research subjects for basic medical research such as space drug testing and disease modeling in orbit. In space, these models can be used to study the effects of space radiation on the human body, record organ function under microgravity, and provide possible solutions for minimizing damage to healthy cells when living or working in space. The microgravity environment in space may also help create larger and more complex engineered tissues. Compared to tissues built on Earth, these tissues look and function more like human tissues.

The European Space Agency has conducted some preliminary research on the 3D printing technology of human tissues needed to keep astronauts healthy during the process of Mars exploration, and has developed a "bio-ink" that can be used for printing skin and bone samples.

Scientists from the Dresden University of Technology Medical School believe that human plasma can be used as "bio-ink" for 3D printing of skin cells (Fig. 8.5). Such "bio-ink" can obviously be obtained from astronauts during space flight. However, with the change of gravity conditions, the "bio-ink" is not easy to adhere to the printed structure, which will prevent the printing work from continuing. Additional materials need to be added to increase the viscosity of the "bio-ink" to adapt to the printing needs under changing gravity conditions. Astronauts can extract these additives from plants and algae, achieving self-sufficiency of 3D printing raw materials during space flight. The 3D printing of bones uses similar "bio-ink" to print human stem cells, and adds calcium phosphate bone binders that are gradually absorbed during the growth process as structural support materials (Fig. 8.6).

Space travel to Mars or other planets will last for several years, and astronauts face many unknown risks. Under the limited space and weight of the spacecraft, it is impossible and impractical to carry enough medical supplies for all unknown possible situations. Correspondingly, 3D bioprinting technology will enable astronauts to deal

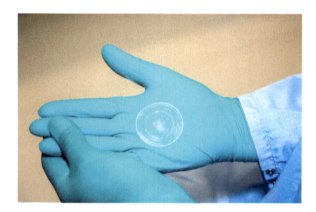

Fig. 8.5 3D printed "bio-ink" sample in space. *Source* European Space Agency

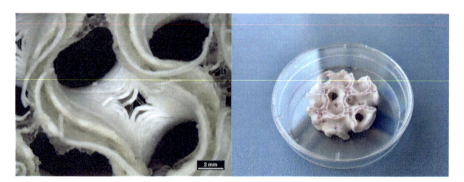

Fig. 8.6 3D printed bone sample. *Source* European Space Agency

8.3 What to Do if an Astronaut Gets Seriously Ill During Interstellar Travel

Fig. 8.7 Russian 3D bioprinter

with medical emergencies that arise. For example, in the case of burns, new skin can be manufactured using 3D printing technology, rather than transplanting from other parts of the astronaut's body, avoiding secondary injuries that may not heal easily in space. And in all cases, the raw materials for bioprinting come from the astronauts themselves, so there will be no transplant rejection issues.

The development of these 3D printed biological samples has taken the first step in the application of 3D bioprinting in space. Subsequent research needs to be carried out on space printing equipment, space operating rooms, and sterile environments.

Currently, Russia has announced the latest progress in 3D bioprinting in space, and Russia has also become the first country in the world to print biological organs in space. Russian astronauts used the 3D bioprinter (Fig. 8.7) on the International Space Station to print the thyroid gland of experimental mice in a zero-gravity environment. Using 3D biotissues in space can study organisms in space to assess the impact of cosmic radiation and other factors. The 3D bioprinter is named Ornanaut and was transported to the International Space Station by the Soyuz MS-11 spacecraft on December 3, 2018.

Furthermore, Russian astronauts have carried out 3D printing of human tissues in a microgravity environment in space, and have manufactured human cartilage from some separated cells with the help of a set of magnetic levitation devices. In traditional human tissue regeneration technology, cells need to be placed on biocompatible support materials. After the cell tissue has completed the printing of the required organs, the support material is degraded to obtain the final human tissue. However, in the experiment of Russian astronauts, using magnetic levitation technology, the cells are suspended in a paramagnetic medium containing gadolinium (a rare earth metal) ions and driven to a specific position by an electric field and voltage.

Currently, the 3D bioprinting is beginning to show promise. However, there are still many scientific problems that urgently need to be researched and solved. There is a long way to go before it can be practically applied in future long-distance space travel.

8.4 3D Printing of Space Landers

Space landers generally refer to instruments that carry out surveying, detection, and cruising missions on the surface of celestial bodies. For the development and utilization of extraterrestrial planets, space landers play a pivotal role. On the surface of the moon, the closest celestial body to Earth, humans have already carried out preliminary exploration work using space landers, uncovering many mysteries of the moon.

From China's Chang'e-2 lander to the still under development Chang'e-7 lander, their mission is to send data and images collected on the moon's surface back to Earth after successfully landing on the moon. They even play the role of "miners", bringing back ores and moon dust obtained from on-site inspections on the moon's surface for data analysis, to help humans better understand the evolution of the moon, and lay the foundation for humans to explore more extensive planets and deep space.

For example, the Chang'e-3 lander carried the Yutu lunar rover, providing necessary hardware support for in-situ detection and patrol detection on the lunar surface (Figs. 8.8 and 8.9). In addition to the lunar rover, landers often coexist with ascenders, orbiters, returners, etc., gradually verifying resource mining, relay communication, sample return functions in one space experiment after another, preparing for the exploration of more distant planets.

The application of 3D printing technology in the construction of space landers has many advantages. On the one hand, the use of 3D printing technology is of great significance for the lightweight design of space landers. To avoid the strong space radiation and extreme environments on the planet's surface, landers often use metals such as aluminum and titanium for construction. The resulting defect is that the lander is too heavy, and the launch and transportation costs increase sharply. However, the use of 3D printing for structural optimization design, especially when combined with AI technology, makes it possible to achieve a design that reduces the weight of the space lander by 10% or even 30%, which has become the focus of research.

Recently, the lightweight design of the lander greatly reduces the cost of sending and space navigation, and also provides the possibility for new types of storage.

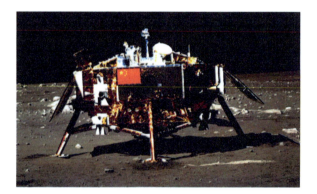

Fig. 8.8 Chang'e-3 lander. *Source* China National Space Administration

8.4 3D Printing of Space Landers

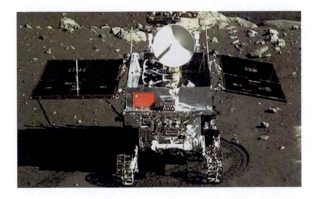

Fig. 8.9 Yutu lunar rover. *Source* China National Space Administration

Figure 8.10 shows the concept of a new type of space lander combining 3D printing technology and AI technology. It has three main structures: an internal structure for placing research instruments printed from aluminum or titanium, a chassis structure that provides support, and metal "legs" realized by 3D printing technology. The bionic concept of a spider is used in the design, allowing the lander to walk easily on the planet's surface. Compared with traditionally designed and processed landers, this spider-shaped lander is much lighter, weighing only about 80 kg. In contrast, the weight of NASA's latest Insight Mars lander reaches 349 kg.

China's Chang'e-4 lunar rover also uses some new parts made by 3D printing. By 3D printing aluminum alloy structural parts, lightweight design is achieved, reducing the weight of the original parts.

Additionally, 3D printing technology also provides the possibility for on-orbit construction and in-situ deployment of space landers. As research continues, it is believed that one day in the future, scientists will be able to bring raw materials to space and print suitable space landers according to actual communication or detection needs.

Fig. 8.10 Concept of 3D printed lightweight space lander

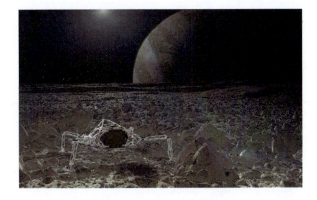

Main Sources of Information

China National Space Administration (CNSA): www.cnsa.gov.cn

National Aeronautics and Space Administration (NASA): www.nasa.gov

Jet Propulsion Laboratory (JPL): www.jpl.nasa.gov

European Space Agency (ESA): www.esa.int

Russian Federal Space Agency: www.roscosmos.ru

Bibliography

1. Xiaoyong T, Dichen L, Bingheng L. Status and prospect of 3D printing technology in space (in Chinese). Manned Spaceflight, 2016,22(4):6.
2. Prater T, Werkheiser N, Ledbetter F, et al. 3D printing in zero G technology demonstration mission: complete experimental results and summary of related material modeling efforts. Int J Adv Manufact Technol, 2019,101:1145.
3. Chu L,Marussi S, Atwood R C, et al. In situ X-ray imaging of defect and molten pool dynamics in laser additive manufacturing. Nat Commun, 2018.
4. Scannapieco D S, Lewandowski J J, Rogers R B, et al. In-situ alloying of GRCop-42 via additive manufacturing: precipitate analysis, NASA/TM-20205003857, 2020.
5. Chen P, Medders M, Katsarelis C, et al.. Segregation evolution and diffusion of titanium in directed energy deposited NASA HR-1, NASA/TM–20210013649, 2021.
6. Hofmann D C, Polit-Casillas R, Roberts S N, et al. Castable bulk metallic glass strain wave gears: towards decreasing the cost of high-performance robotics. Sci Rep, 2016,6:37773.
7. Anon. The first 3D printed Russian satellite will be launched on August 17th. Spacecraft Recov Remote Sens, 2017,38(4):1.
8. Jie Y, Jing L,Wenjie W, et al. Research status and prospect of on-orbit additive manufacturing technology for large space truss (in Chinese). Mater Rep, 2021,35(3):03159.
9. Binbin G. Comparison of the first space 3D printing between China and the United States (in Chinese). Dual Use Technol Products, 2020,8:5.
10. Jingjing L, Yanhong Y, Tao J, et al. Research status of 3D printing technology for metals in space (in Chinese). Manned Spaceflight, 2017,23(5):7.
11. Jizhou G, Qiwen D. Analysis on development situation and environment conditions of our 3D printing technology (in Chinese). Natl Defen Sci Technol, 2015,36:(3).
12. Ping J, Hui L, Zongtan S. Space application of 3D printing in foreign countries (in Chinese). Space Int, 2015,4:4.
13. Mcguire T, Hirsch M, Parsons M, et al. Design for an in-space 3D printer. Sens Syst Space Appl IX, 2016.
14. Analysis of specimens from phase I of the 3D Printing in Zero G Technology demonstration mission. Rapid Prototyp J, 2016.
15. Zou Y, Li W. China's deep-space exploration to 2030 (in Chinese). Chin J Space Sci, 2014,5(34):516–7.
16. Hofmann D C, et al. Compositionally graded metals: a new frontier of additive manufacturing. J Mater Res, 2014,29(17):1899–1910.
17. Feng L. The solution of future space transport by three-dimensional printing technology (in Chinese). Tactical Missile Technol, 2013,6:5–9.
18. Jianping Z. Chinese space station project over all vision (in Chinese). Manned Spaceflight, 2013,19(2):1–10.

19. Li M, Luju H, Xianyi S, et al. Micromorphology and microstructure stability of TiAl alloy deposited by electronic beam (in Chinese). J Mater Eng, 2016,44(1):89–95.
20. Cesaretti G, Dini E, Kestelier X D, et al. Building components for an outpost on the Lunar soil by means of a novel 3D printing technology. Acta Astronaut, 2014,93:430–450.
21. Hoyt R, Cushing J, Slostad J. SpiderFab™: process for on-orbit construction of kilometer scale apertures. NASA Innov Adv Concepts (NIAC), 2013.
22. Gradl P R, Protz C, Fikes J, et al. Lightweight thrust chamber assemblies using multi-alloy additive manufacturing and composite overwrap. In: AIAA propulsion and energy 2020 forum. 2020.
23. Martin-Iglesias P, Vorst M, Gumpinger J, et al. ESA's recent developments in the field of 3D-printed RF/microwave hardware. In: Proceedings of the 2017 11th European Conference on antennas and propagation (EUCAP). IEEE, 2017.
24. Yifei L, Liang L, Gong W, et al. Application of metal additive manufacturing technology for space (in Chinese). Chin J Space Sci, 2018,38(3):6.
25. Neil L, Weiran Z. 3D printing in Space (in Chinese). Green Low-Carbon Technol Ind, 2020,8:5.
26. Jiefu S, Yan S, Xiaosong Q, et al. Design of new 3D printing broadband dielectric resonant navigation antenna (in Chinese). In: China Satellite Navigation Conference, 2019.
27. Zheng Y C, Wang S J, Li C L, et al. The development of CAS-1 lunar soil simulant. In: Proceedings of the International Lunar Conference, 2005.
28. Hafley R, Taminger K, Bird R. Electron beam freeform fabrication in the space environment. In: AIAA Aerospace Sciences Meeting and Exhibit, 2007.
29. Fiske MR, Mcgregor W, Pope R, et al. Lunar in situ materials-based surface structure technology development efforts at NASA/MSFC. In: AIP conference proceedings, 2007.